D0728236

RAPE
Weapon of Terror

RAPE
Weapon of Terror

Sharon Frederick
and
The Aware Committee on Rape

Association of Women for Action and Research

HV
6558
.F74
2001
C. 1

Published by

Global Publishing Co. Inc.
1060 Main Street, River Edge
NJ 07661, USA

RAPE: WEAPON OF TERROR

Copyright © 2001 by Global Publishing Co. Inc.

ISBN   1-879771-53-5

$12.00

Printed in Singapore by World Scientific Printers

This book is dedicated to all victims of rape past, present and future. It is also dedicated to the memory of Dorothy Cheung, who helped inspire the writing of this book as a founding member of the AWARE committee on rape.

FOREWORD

The human, social and economic cost of gender-based violence is incalculable. Transcending boundaries of nationality, culture, religion and socioeconomic status, violence against women and girls in its different manifestations plagues every society through harmful consequences to women's physical and emotional health, loss of self-esteem, and as a leading cause of death of women through murder and suicide. Affecting the capacity of over half the world's population to freely and fully participate in the development process, violence against women and girls carries exorbitant social and economic costs that have yet to be fully and accurately assessed.

The issue of violence against women has brought some of the century's greatest challenges, but also some of the most powerful victories for human rights that have served to place violence against women onto the international agenda as a human rights and a development concern of the highest priority. The United Nations Development Fund for Women (UNIFEM) supports innovative initiatives worldwide to combat all forms of violence against women, including domestic violence, sexual violence and harassment in the school and in the workplace, rape and sexual slavery in the context of war, trafficking in women across borders, as well as violence associated with traditional practices.

I commend the Association of Women for Action and Research (AWARE) for their initiative to publish the book *Rape: Weapon of Terror*. This book promises to serve as a tool for advancing the goal of eliminating violence against women, especially in the context of situations of war and civil conflict. The use of gender-based violence, including rape and forced pregnancy, is increasingly a horrifying feature of war in ethnic conflicts. As more and more women and girls are abducted by warring groups, displaced from their homes and communities, and threatened with deadly diseases such as HIV/AIDS, they are at the same time called upon to assume wider responsibility for holding their communities together, taking new leadership roles and often sustaining their families single-handedly. All of these experiences have brought home to many women the vivid links between violence, poverty and inequality. They see clearly the continuum of conflict that stretches from the beating at home to the rape on the street to the killing on the battlefield.

This new book serves to broaden our knowledge and understanding of gender-based violence as a weapon of war. At the same time it also highlights the need for effective intervention in the area of peace building and conflict resolution. There is a need to find common ground between civil societies and states, the United Nations system and business communities so as to rebuild institutions and capacities, and to bring international agreements and treaties to bear upon the operation of economic and political systems.

Noeleen Heyzer
Executive Director, UNIFEM

CONTENTS

RAPE

Weapon of Terror

INTRODUCTION

Why did you pick up this book?

Were you intrigued by its provocative title? Did a friend or colleague recommend it? Are you simply curious about the topic, about which you know little?

Whatever your reason, the more of this book you read, the more you will be tempted to deny the information it contains, or its relevance to you.

If you are a woman, you may wince at the descriptions of violence done to other women, but distance yourself by concluding such things happen only in less fortunate places, to far less fortunate women. If you are a man, you may question the statistics, assuming they are inflated, and conclude that any book on rape is probably a feminist tirade.

Alternatively, you may accept the validity of the data, but put the book aside because you feel frustrated and powerless. Why read about horrible actions that you can do nothing to stop? What good does it do? Why upset yourself?

These reactions are natural and typical of how most people respond to new knowledge about the suffering of others. They are typical — and they are also one of the most powerful reasons why such suffering continues.

There is another, more helpful and satisfying way to respond to the distressing information in the pages that follow. Read with an open mind; be prepared to change your view if the information is credible and convincing; allow yourself to feel empathy for the victims of violence, to feel the distress and anger that can lead to action.

Understanding and action are the goals of AWARE, Association of Women for Action and Research, the Singapore-based organization which conceptualized and sponsored this volume. The book aims to give the lay reader accurate information on a complex, emotion-laden topic as well as the understanding and motivation to take action — actions like those described in the final chapter.

The idea for the book sprang from the hearts and minds of a small group of women active in the work of AWARE as they watched stories of sexual assault unfold in Bosnia, Rwanda, and then very close to home, in post-Suharto Indonesia. They first mounted a public information campaign, including an exhibition on mass rape as a weapon of terror. The crowds which visited the exhibition were clear evidence of the tremendous interest in the topic. Equally clear was the need for a more permanent and complete discussion of the issue — thus, the idea for a book was borne.

Stripped to its statistical skeleton, the story is both clear and ugly. A few examples:

- World War II documents, the best recorded evidence of wartime rape, reveal assaults numbering at least several hundred thousand, perhaps as many as two million: Thousands in the villages of Russia and Poland, as the Germans invaded early in the war; thousands more when the Soviets got the upper hand and took revenge on the bodies of German women. In the final two weeks of the war, an estimated 100,000[1] German women were raped in Berlin, by victorious Russian and other Allied troops. In Asia, figures are more exact: at least 20,000 in the Chinese wartime capital of Nanking when the Japanese invaded China; at least 80,000 — perhaps over 100,000[2] — Korean, Indonesian, Filipino and Chinese women repeatedly raped during their months as sex slaves of the Japanese soldiers.
- In the decades that followed World War II, the international community paid little attention to, and therefore did little to

document, rape during armed conflict though we know a significant number of assaults occurred in areas such as the Congo, Peru, El Salvador, Cambodia and Vietnam.[3] It is only in 1972 that we have clear evidence of a reign of terror-by-rape rivaling World War II atrocities. When Bengal (officially East Pakistan) declared itself the independent state of Bangladesh, West Pakistani troops quickly moved in to quell the rebellion, and to terrorize the population of 75 million by carrying out widespread rape and murder. During the nine-month war that followed, 200,000[4] Bengali women (a conservative estimate) were raped in their homes, on the streets, and in military barracks where many were kept as prisoners for the nightly use of the men. At least 25,000 pregnancies[5] resulted from the assaults. Asked why mass rape had been used systematically in Bangladesh, a Bengali politician responded "Put a gun in (soldiers') hands and tell them to go out and frighten the wits out of a population and what will be the first thing that leaps to their mind?"[6]

- During the last decade, rape as a weapon of terror has been documented by news media and international aid organizations in countries including Afghanistan, Kuwait, Algeria, Indonesia, Somalia, Haiti, Kashmir, and Sierra Leone. In the most notorious incidents, more than 20,000 women and girls were raped between 1992 and 1994 as part of the so-called "ethnic cleansing" in the Balkans. An estimated 200,000 to 400,000 women were raped in Rwanda during the genocidal 1994 war that killed between 500,000 and one million people.

What is the character of these rapes? Is rape during armed conflict different from that which occurs under normal conditions?

Some may argue that wartime rape is an inevitable outcome of combat, an outlet for men separated from normal female companionship and subjected to deprivation and danger. The evidence does not support that view. Even some military commanders, speaking off the record, acknowledge that the presence of women — for example, in semi-official brothels — does not affect sexual assaults on "enemy" women. "Rape has nothing to do with the availability of willing women and prostitutes," acknowledged one member of the US military hierarchy.[7]

More importantly, the circumstances of the rapes themselves argue against rape as an outlet for sexual frustration, to be explained with a shrug and a shake of the head since "men will be men". Most rapes are extremely cruel, often marked by sexual mutilation and even death. Rape during times of conflict is "... the expression of rage, violence and dominance over a woman."[8] It is "... virtually always about power and contempt. Virtually always the effect...is humiliation, degradation, subordination, and severe physical or psychological injury...".[9]

Civilian targets of war

In fact, the world today is a far more dangerous place for all civilians than ever before as "more and more of the world is being sucked into a desolate moral vacuum in which civilians rather than soldiers are the main targets in war".[10]

No longer are the main terrors "the big wars", those of major powers setting their nuclear arsenals against one another. Nor are they the wars waged on clearly defined battlefields between trained armies of soldiers. They are more often wars of ideology, fought by militias, roving bands of angry tribes, child soldiers, former neighbours out to right perceived ethnic wrongs. Just one telling statistic: In World War II, one of every two casualties was a civilian; in the last 50 years, it is estimated that armed conflicts kill nine civilians for every one combatant.[11]

These civilians are both men and women, adults and children, of all races, religions, ethnic groups and social status. But when it comes to sexual assaults, the overwhelming majority of victims are female.

"...the same atrocities which happen to the civilian male happen to the civilian female: both men and women are shot, burned, bayoneted, hung, beaten, bombed, tortured, forced into slave labour."

However, women face additional atrocities since "...females are sexually assaulted with alarming regularity...

"Throughout the history of war, while male civilians are killed, female civilians typically are raped, then killed. In torturous interrogations, males are savagely beaten. Females are savagely beaten and raped. Conclusively, all civilians are not treated similarly, although the law groups them into one general category. This law that applies to all civilians has tended not to recognize the sexual abuses routinely committed against over half of the civilian population — the women."[12]

Law is based on custom and precedent and, from earliest times, rape has been seen as an inevitable and even sanctioned behaviour during armed conflicts. Virtually all early civilizations viewed women as chattel, property owned by men just like cattle and grain, and therefore just as vulnerable to being "taken" by the victor.

Even as states became more civilized, with codes and laws developed to mitigate the horrors of war, rape was seldom punished as a crime. More often than not, soldiers on the winning side felt they were "owed" the women of the conquered nation, partly as booty, partly as a final symbolic expression of their victory and their enemy's humiliation.

In the modern world, rape has become more malignant, used as a stated strategy for terrorizing an entire civilian population, either to subjugate them to the will of the attackers as in World War II; or to physically displace or annihilate them as a people, as in Bosnia and Rwanda. It has became a way of destroying a nation or group both physically (many brutally raped women can no longer bear children) and culturally since women play a central role in family and community structures. Breaking down the women is a very effective method for breaking down the community.

Why are these horrors allowed to happen, often with impunity? Why don't other nations intervene when hundreds and thousands of women are assaulted as they have been as recently as the last decade, as they are right now?

Rape in the eyes of the law

Subsequent chapters will attempt to provide some responses to these questions. Part of the answer, though, lies in the way rape has been viewed by the law. Because armed conflicts throughout the ages have had such a horrific impact on societies, nations have developed agreements over time to limit the destruction: to determine how wars could be fairly fought and how noncombatants should be treated. (Such codes of behaviour, for example, date from as early as the classic *Art of War* written by Chinese warrior Sun Tzu in 700 BC and are even more fully developed in the Hindu Code of Manu from 500 BC.)

Today the body of law governing armed conflict — based largely on treaties and custom — is labeled "international humanitarian law" or the "law of war". Acts committed during armed conflict or "war crimes" are violations of the law of war that warrant prosecution of individuals.

Such laws have tended to look at rape as a crime against honour. In earlier centuries it was the woman's father, husband or guardian who was "dishonoured" most by a rape since she, his property, had lost significant value. In more recent times, rape is seen as dishonouring the woman and her family.

Even the Geneva Conventions of 1949, an agreement forged out of the horrors of World War II and a major source of modern humanitarian law, reflects this point of view. It includes rape among those acts which constitute an attack on a woman's honour, but fails to mention rape in its discussion of a person's right to physical integrity.

This is a dangerous line of reasoning. "The pitfalls in linking rape and honour are many. First, reality and the woman's true injury are sacrificed...violations of honour and modesty are wholly inadequate concepts to express the suffering of women raped during war... Second, by presenting honour as the interest to be protected, the injury is defined from...(the community's) viewpoint," thus promoting...the notion that the raped woman is soiled or disgraced...

Third, on the scale of wartime violence, rape as a mere injury to honour or reputation appears less worthy of prosecution than injuries to the person. The failure to recognize the violent nature of rape is one reason that it has been assigned a secondary status in IHL" (international humanitarian law).[13]

In addition, neither the Geneva Convention, nor its additional protocol of 1977, define rape as one of a specific set of war crimes that it labels "grave breaches". This is important because if an action is defined as a grave breach, a nation is required to prosecute persons responsible or hand them over to a nation which will do so.

Only recently, in the 1993 charter of the UN tribunal for the former Yugoslavia, was rape finally defined as a so-called "crime against humanity", that is, one of a number of very serious inhumane acts committed as part of a systematic attack against a civilian population. This category of crimes was established by the World War II Nuremberg Tribunal, in the wake of Nazi horrors. So egregious are these crimes that most nations accept a responsibility to intervene, even if a government is taking such actions against its own citizens. (This is a direct contradiction of earlier international law which essentially held that a nation was entitled to treat its own citizens as it chose.)

And, finally, laws or no laws, rape continues because "Soldiers have not feared punishment for sexual violence toward a part of the civilian population that, even in times of peace, held an unequal or diminished status vis à vis men."[14]

As a result of many historical, social and legal conditions, wartime rape has been consistently misunderstood, mislabeled, downplayed, or outright ignored. That is beginning to change: for example, the United Nations Tribunals for the former Yugoslavia and Rwanda are breaking new ground in their prosecution of mass rapes. But the process is painfully slow...and often woefully misunderstood.

The chapters that follow attempt to speed up the process of understanding rape used as a weapon of terror — whether as part of a declared war between nations, an internal battle between political or ethnic groups, or a government's actions against its own citizens. Chapter 1 looks at the historical underpinnings of the problem. Chapters 2, 3 and 4 review the sad record of rape during the last decade, first in the Americas, Europe and Africa, then in Asia — a special focus of this book. The final chapter provides a summary of key issues and examines the current state of legal and political efforts to stop sexual violence. It also provides practical steps that you as an individual can take to promote understanding and action, to help bring justice to the thousands of women who have been sexually assaulted as part of armed conflicts.

A final thought: Rape is just one of the many terrible ways in which we human beings abuse one another. You need not be female or a feminist to deplore this particular violation of body and spirit. You only need to be a believer in that most basic of human rights, the right of every person to control the integrity and privacy of his or her own body.

Notes

[1]Hilkka Pietila and Jeanne Vickers, *Making Women Matter, The Role of the United Nations* (1994), p. 146, as cited by Kelly Dawn Askin in *War Crimes Against Women* (1997), p. 52.

[2]George Hicks, *The Comfort Women* (1995), p. xix.

[3]According to veteran war reporter Peter Arnett, who covered the Vietnam War, the Vietcong and North Vietnamese seldom committed rape. It was a serious crime, punishable by military execution. Reasons he cited included the Vietcong's dedication to what they saw as a revolutionary mission and the fact that Vietcong women fought as equals in military operations. Author Susan Brownmiller who quotes Arnett in her 1975 classic on rape, *Against Our Will: Men, Women and Rape*, pp. 90–91, goes on to say: "The concept that a revolutionary guerrilla army of peasants does not rape was laid down with simple eloquence in 1928 by the great Chinese general Chu The, whose motto was 'Take not even a needle or thread from the people.' His rules included a clear directive: 'never molest women'."

[4]Brownmiller, pp. 79–80.

[5]Brownmiller, p. 84.

[6]Brownmiller, p. 85.

[7]Brownmiller, p. 76.

[8]Ruth Seifert, "War and Rape: A Preliminary Analysis", p. 55, in *Mass Rape: The War Against Women in Bosnia-Herzegovina*, edited by Alexandra Stiglmayer (1994).

[9]Askin, p. 16. In fact, most feminists would argue that rape, inside or outside of war, is an act of violence, not lust, and results from society's view of women. For example, Catherine Morrison argues that "rape is the ultimate extension of our culture's normal tendency to regard women as inferior to men, as related to men in a useful or objective way, as servants, as possessions, or as badges of honour". Arguments such as these are beyond the purview of this book, but interested readers can find further discussion in Brownmiller's book and Seifert's article, cited above, and in "A Cultural Perspective on Rape", by Catherine Morrison, in the *Rape Crisis Handbook* (1980), pp. 3–16.

[10]*Impact of Armed Conflict on Children*, p. 1, Report of Graca Michel, Expert of the Secretary-General of the United Nations, UNICEF, 1995.

[11]Roy Gutman and David Rieff, *Crimes of War: What the Public Should Know* (1999), p. 10.

[12]Askin, pp. 12–13.

[13]Catherine N. Niarchos, *Women, War and Rape: Challenges Facing the International Tribunal for the Former Yugoslavia*, pp. 674–675, Human Rights Quarterly, Vol. 17, No. 4, November 1995, The Johns Hopkins University Press.

[14]"Violence Against Women", p. 5, in *World Report 1999*, Human Rights Watch.

LESSONS FROM THE PAST

"A person born a female had rights over nothing: not their own bodies, their own sexual integrity."[1]

Catherine Niarchos, *Human Rights Quarterly*

In 1474 Peter van Hagenbach, a knight and military officer, ensured his place in history when he took the witness stand as the first individual ever to be tried for rape by an international tribunal. In his case, he faced 27 judges of the Holy Roman Empire, representing Swiss, Alsatian and German states.

Hagenbach was found guilty of crimes of rape (as well as pillage and murder) committed by his troops during a brutal occupation of the Austrian town of Briesbach. His sentence called for beheading.

Yet Hagenbach would never had been convicted if he had been waging war, rather than conducting a military occupation. If he had declared war and thus brought the so-called rules of war into play, he would likely have been judged hero rather than criminal. For in the Middle Ages in Europe, the rules of combat allowed soldiers to kill at will provided a war was in progress. And, "in the case of cities which had refused to surrender when surrender had been demanded, the women occupants could be raped".[2]

The history of armed conflict — which always brings with it sexual assault — is one of a slow march from outright, no-holds-barred violence to hard-won agreements among warring states. Almost all early states came to believe that some rules should apply to how soldiers behaved both to other combatants and to the civilian population. In the Western world, the writings of people like Aristotle, Cicero, St. Augustine, and St. Thomas Aquinas established the guidelines for waging a "justified" war. In the Islamic world, the Koran established certain codes.[3] There were also "agreements in place between the city-states of classical Greece and between the Hindu kingdoms of ancient India".[4]

In general, these ancient legal systems were based on two principles: Anything done to defeat the enemy was sanctioned; anything that was not directly related to the enemy's defeat and caused unnecessary suffering was not sanctioned. Defining which was which, however, was seldom clear — except when it came to women. In every state, "A person born a female had rights over nothing: not their own bodies, their own sexual integrity."[5] Therefore, females were fair game in war.

For example, "Among the ancient Greeks, rape was socially acceptable behaviour well within the rules of warfare, an act without stigma for warriors who viewed the women they conquered as legitimate booty, useful as wives, concubines, slave labour or battle-camp trophy."[6]

By the fourteenth century, as philosophical and religious thought developed in both East and West, there were enlightened thinkers who argued that all civilians should be protected during armed conflict — including protection from rape. "Soldiers may not be given license to murder, rob, plunder, rape or constrain civilians; those who do such things should be as severely punished as if the crimes had been committed in peacetime," argued legal scholar Lucas de Penna.[7]

But three centuries later, such admonitions against rape had still not been widely heeded. In 1646 when Hugo Grotius, the father of modern international law, compiled current law and practices into a code of wartime behaviour, he acknowledged that:

> "Those who sanction rape... have judged that it is not inconsistent with the law of war that everything which

belongs to the enemy (including the women) should be at the disposition of the victor. A better conclusion has been reached by others...and consequently (rape) should not go unpunished in war any more than in peace."[8]

Women as property, banner and booty

In fact, from ancient times through the sixteenth century, powerful undercurrents argued against more "civilized" and humane treatment of women during war.

One of the most powerful undercurrents was the assumption that every woman was considered the lawful property of a man: a father, husband, guardian or master. Therefore, even as systems of law developed, rape was defined as a property crime against the man who owned the raped woman.

"The ancient perspective of women as spoils of war was consistent with the legal status of women. Women were regarded as virtual property — certainly they were without legal capacity — and rape committed by a member of the community was consequently an injury to the male estate and to the community, but not to the woman."[9]

> "The crime (of rape) was principally that of stealing or abducting a women from her rightful proprietors, normally her father or husband...Moreover, in the case of a maiden, rape destroyed her property value on the marriage market, and... heaped shame on her family. Violated daughters might be given as offerings to nunneries, and in many societies they were married off to the abductor or rapist."[10]

In addition, from earliest times soldiers have seen the rape of enemy women as the final humiliation of their adversaries — and their adversaries have agreed. "Defense of women has long been a hallmark of masculine pride, as possession of women has been a hallmark of masculine success. Rape by a conquering soldier destroys all remaining illusions of power..."[11]

During the middle ages, soldiers also believed that they were "owed" opportunities for rape and pillage. These activities were seen as compensation for the ordinary soldier's own hardships, including poor, irregular pay, during long campaigns and siege warfare. The license to rape was "a tangible reward for services rendered...(and) an actual reward of war".[12]

However, we have little evidence to suggest that military commanders actually *ordered* soldiers to commit acts of rape.[13] While women may have been treated brutally over the centuries, the brutality stemmed from assumptions about the place of women and the role of the victorious warrior — it was not an official strategy of war. That would come later.

Foundations for today: Rape as a crime against honour

Our modern system of governing relations between states, known as international law, dates from the 16th and 17th centuries and parallels the development of modern European political states. Non-European states had their own systems of law, but these were absorbed into that of Western Europe as the Europeans colonized much of the rest of the world. As they gained independence, non-European states largely accepted the Western-led legal system, working to revise those parts of it which they believed were contrary to their interests.[14]

The foundation of contemporary thinking on sexual assault was built during the 1800's; a series of national codes and international agreements developed as nations began to recognize the increasing horrors of modern warfare. Some commentators see the rise of wartime press coverage, thanks to the invention of the telegraph, as a critical factor. For the first time, ordinary people could read and hear distressing stories about the pain and suffering of civilians, of people like themselves. They were no longer limited to carefully screened tales of bravery based on "the often self-serving accounts of military commanders long after the events".[15]

In conflicts during the second half of the century — for example, the Crimean War in Europe (1854–56) and the War between the States in America (1861–65) — the public began to read about the horrifying reality of combat. American president Abraham Lincoln reacted to the war coverage by asking law professor Francis Lieber to draft a comprehensive code by which land wars should be fought.

The so-called Lieber Code became an important cornerstone of international law, and of laws protecting women against assault. But, it reflects an unfortunate ambiguity borne out of old assumptions about women as property. In one provision, the Lieber Code gave special protection to women because of their role within families. In another, it simply and forcefully states that all rape is prohibited under the penalty of death. Thus, we are left wondering "whether (rape) is a crime against women, or against men and community".[16]

In 1899 and again in 1907, peace conferences were convened at The Hague in Belgium. The results were 13 conventions or agreements, which codified generally accepted principles of international law, covering prisoners of war, treatment of civilians, and methods of warfare. With regard to the treatment of women, the 1907 Hague Convention IV is the most relevant: "... family honour and rights... must be respected". Article IV thus encourages the view that assaults on women are an attack on the family, not violent criminal acts against individuals.

World War I: Rape as a tool of terror and propaganda

These were the primary agreements in effect when war broke out in Europe in 1914. Despite them, World War I was the bloodiest in history, with approximately the same number of lives lost (5 million) in four years, as had been lost during all wars in the previous century. For women in Belgium and France, the war was a worse nightmare than they could ever have imagined.

In a particularly vicious three-month invasion that marked the beginning of the war, German soldiers terrorized village after village first in Belgium, then in France. They burned houses, raped women, killed inhabitants. An eyewitness who survived the plunder of one of these towns, Louvain in Belgium, described the scene: "The women and children were separated... Some German soldiers came up to me sniggering and said that all the women were going to be raped.... They explained themselves by gestures."[17]

These rapes were not simply "indiscriminate acts committed by the occasional soldier on his own initiative. They were used as a weapon of terror, rage, and intimidation...".[18] In Belgium, it was reported that 'Outrages upon the honour of women by German soldiers have been so frequent that it is impossible to escape the conviction that they have been condoned and indeed encouraged by German officers'."[19] Thus, for the first time, we have evidence of mass rape being used quite deliberately as a weapon to intimidate and terrorize an enemy.

Rape is also used quite clearly and deliberately as a tool of propaganda. Lurid reports on the brutality of the "Huns", including the rape of innocent women, were used to build anti-German feelings and gain public support for the war. "As propaganda, rape was remarkably effective, more effective than the original terror."[20] However, once the tide of war turned in favour of the Allies, the propaganda mills slowed and public attention turned elsewhere. Initial public outrage was largely forgotten by the time of the armistice in 1918.

An international war crimes commission was formed in 1919 to report on German atrocities; rape and forced prostitution were two of the 32 offences specified. Unfortunately, squabbling among the Allied powers meant that no tribunal was ever convened to charge German soldiers with war crimes. In a compromise move, the German Supreme Court was allowed to try those accused of war crimes. Of 901 men charged, only 13 were convicted and those 13 never served their prison terms.[21]

World War II in Europe: Establishing a master race

With no punishment for the atrocities of the First World War, and thus no deterrent for the future, troops in World War II engaged in even more blatant assaults. It was the most horrific example of sexual assault the world had yet seen, driven by the ideologies of the Nazi and Imperial Japanese aggressors. "Rape for the Germans, and to a similar extent for the Japanese, played a serious and logical role in the achievement of what they saw as their ultimate objective: the total humiliation and destruction of 'inferior peoples' and the establishment of their own master race."[22]

In Europe, the Germans assaulted the female population of virtually every population they invaded.[23] The assaults were so widespread that it is generally acknowledged that they must have been at least tacitly approved by Nazi officers, perhaps even ordered. In Poland, for example, an affidavit submitted at the Nuremberg Trials includes a description of the attack on the town of Lvov:

> "...32 women workers of the Lvov clothing factory were raped and then killed by German storm troopers... Everywhere the bestial German bandits break into houses, rape women and girls before the eyes of their relatives and children, torture their victims and brutally murder them on the spot...

"Reports of the vile outrages committed against women and girls, schoolgirls and small children by the Germans during their occupation are pouring in daily from villages and towns recently liberated..."[24]

Another report, also from Poland:

> ... the Fascists exceeded all boundaries of human imagination. Before the eyes of mothers, who were either fainting or losing their wits, the drunken German soldiers and policemen shamelessly raped the girls before each other or others. They cut out the sexual organs with daggers,

forced living and dead bodies to assume the most disgusting poses, cut off noses, breasts, and ears."[25]

As Germans entered the Soviet Union, similar assaults were widespread; in fact, evidence indicates that "in some towns and villages every woman and girl was subjected to sexual assault. Every single one. This indicates an organized and systematic plan to rape and further destroy the women inhabitants".[26]

Sexual assault was not limited to Germans, however; in fact, all parties to the war committed crimes of rape. For example, Moroccan mercenaries fighting as part of the French force in Italy in 1943 raped great numbers of Italian women, having been given tacit permission to "have their way" with the local population. This was no different than the behaviour of commanders and troops from centuries before, when women were clearly viewed as property to be used as men chose.[27]

When the war turned in favour of the Allies, Russian armies revenged the brutalities of the Germans by brutally assaulting "their" women. "These German women had not marched into Russia and committed horrible crimes; but nevertheless, the German women and girls were victimized because of their nationality, because of their relationship to the male soldier, and because of their gender."[28]

Author Alexander Solzhenitsyn, then an officer in the Russian army, acknowledged: "All of us knew 'very well that if the girls were German they could be raped and then shot. This was almost a combat distinction."[29]

The result of thousands of sexual assaults across Europe? Although the victorious Allied nations established the landmark International Military Tribunal at Nuremberg to try European war criminals, the Tribunal essentially ignored crimes against women. This was in spite of the fact that Tribunal trial transcripts "contain evidence of vile and torturous rape, forced prostitution, forced sterilization, forced abortion, pornography, sexual mutilation, and sexual sadism".[30]

There were no prosecutions, nor was there any effort to at least publicly document sexual atrocities. Quite the contrary. The French prosecutor, for example, actually told the court, with some air of gallantry, that he would avoid citing the awful details of the rapes which had been committed.

Thus, "Whether it was out of shyness, prudishness, reserve, ignorance, revulsion, confusion, or intentional omission, the lack of both public documentation and official prosecution gave impetus to the notion that sexual assaults were less important crimes...".

"The gravity of the sexual assaults (were) minimized by the failure of the international community to confirm their legal and moral importance."[31]

World War II in Asia: Widespread sexual atrocities

In the Asian theatre of war, the picture of atrocities committed against women was even grimmer than in Europe.

Foremost among these atrocities is the infamous "Rape of Nanking" by Japanese troops who entered China's wartime capital in December 1937.

The first three months of Japanese occupation were characterized by extremely brutal and widespread sexual assault. Twenty thousand rapes were documented by the Tokyo Tribunal as occurring during the first month of occupation. Many observers believe the total number of women raped was as high as 80,000.

Even the word "rape" seems too mild a term to describe what occurred in Nanking. Brutal assaults took place at every hour of the day and night, and in every place in the city — in the victim's home, in full view of family members; on the streets, watched by horrified passers-by; in hospitals, where neither nurses nor patients were safe; in schools and university buildings; in churches; and in declared safety zones.

No one was safe — not old women, or young girls, or pregnant mothers. "Many women in their eighties were raped to death...Little girls were raped so brutally that some could not walk for weeks afterwards...The Japanese violated many who were about to go into labour, were in labour or had given birth only a few days earlier."[32]

"Perhaps one of the most brutal forms of Japanese entertainment was the impalement of vaginas. In the streets of Nanking, corpses of women lay with their legs splayed open, their orifices pierced by wooden rods, twigs and weeds...The Japanese drew sadistic pleasure in forcing Chinese men to commit incest — fathers to rape their own daughters, brothers their sisters, sons their mothers."[33]

Nor can the behaviour of the Japanese army be explained as the acts of soldiers who went out of control for a brief time, after a hard battle. In fact, troops committed brutal sexual assaults, as well as arson and murder — all on a large scale — for at least six weeks after they took Nanking.[34]

Large numbers of sexual assaults took place in other cities taken by the Japanese. In Manila, for example, "[Japanese Admiral] Iwabuchi issued further orders...lay waste to the city!...Beer, sake, and wine from Manila's shops were distributed to the men, and within hours the 20,000 man force was a raging, drunken mob...In an intoxicated fury of revenge and despair, the wild-eyed sailors threw themselves into an orgy of burning, shooting, raping and torture. Young girls and old women were raped and then beheaded...".[35]

International outrage ran high as news of the "Rape of Nanking", including the massacre of more than 260,000 civilians, made its way to the Western world. In response, the Japanese military made plans to limit future assaults on civilian women by initiating a large-scale, multi-national "comfort system" — a secret enterprise that represented "the legalized military rape of subject women on a scale — and over a period of time — previously unknown in history."[36] At least 80,000 women from Japanese-occupied territories — Korea, the Philippines, Indonesia, China, Taiwan, Malaysia, and Burma — were either lured or forced into the comfort system to provide "sexual services" to the Japanese military for the remainder of the war. (The largest group, an estimated 80 percent, were Koreans.)

When the war ended, Japanese leaders of the military effort were prosecuted for "crimes against peace" — that is, the crimes associated with waging an aggressive war — during the International Military Tribunal for the Far East (IMTFE) which opened in Tokyo in May 1946, a few months before the Nuremberg Tribunal rendered its verdicts. The trial lasted two and a half years and charges were brought against 28 defendants, half of whom were generals in the Imperial Japanese army, all of whom were judged to have been leaders with major responsibility.

In Tokyo, unlike Nuremberg, rape was one of the war crimes with which some of the defendants were charged and prosecuted, citing as evidence the assaults on more than 20,000 women and girls by Japanese troops in Nanking. Rape was charged in the indictment as a war crime, under "inhuman treatment", "ill treatment" and "failure to respect family honour and rights", establishing precedent for prosecuting rape as a war crime. Based on all the charges against them, all 28 defendants were found guilty.[37]

In another important post-World War II trial, separate from the Tribunal, General Tomoyuki Yamashita, commander of the 14th Area Army, was accused not of waging a war of aggression — "the litmus test for trial before the IMTFE"[38] — but with traditional war crimes. He was prosecuted for crimes, including murder and rape, committed by his troops during ruthless assaults on thousands of Filipinos.

"In whispers and in screams, (the court) heard how over 32,000 Filipino citizens had died...It learned of rape and necrophilia: of how 476 women in Manila were imprisoned in two hotels and repeatedly raped over an 8-day period by officers and enlisted men alike; of how twenty Japanese soldiers raped one girl and then, as a grand finale, cut off her breasts; and of how drunken soldiers, after killing women civilians, then raped the corpses."[39]

For the first time in modern history, a commanding officer was held criminally liable for acts committed by his troops. He was sentenced to death. "...Yamashita's trial is important because it held that, in certain circumstances, a superior can be held accountable for, among other things, rape committed by persons under their command."[40] Thus, it established a precedent "for prosecuting leaders or commanders for gender specific crimes committed by persons under their authority".[41]

As we look back through today's lens, it seems impossible that the horrific sexual assaults of World War II were almost forgotten. How could it have happened? From the distance of half a century, a recent United Nations report reflects on some of the reasons.

"Part of the problem is that sexual violence was perpetrated by all sides to the conflict. Consequently, it was difficult for one party to make allegations against the other at the conclusion of hostilities. Moreover, sexual violence had long been accepted as an inevitable, albeit unfortunate, reality of armed conflict. This was compounded by the fact that in the late 1940's sexual matters were not discussed easily or openly, and there was no strong, mobilized women's movement to exert pressures for redress."[42]

In addition, "Perhaps prosecutions of crimes violating the right to life (murder, reprisals, hostage-taking) were considered more weighty. The more likely explanation is that rape as a *distinct* category of atrocity was not recognized; it was not seen as persecution based on gender. It may be anachronistic to expect the... tribunals to have grasped this concept; after all, it is still considered novel today."[43]

Whatever the reasons, once again rape was a forgotten war crime that went virtually unnoticed and virtually unpunished. As a result, rape continued to be a weapon of choice for conflicts that followed: in El Salvador, Sri Lanka, Uganda, Peru, Kuwait, Somalia, Vietnam, Cypus, Sudan, Cambodia, Bangladesh, Algeria. In 1972, the United Nations sent out a warning that went largely unheard and unheeded: war cruelty, especially toward women, continued at virtually the same levels despite the international agreements that were in place. And, so the quiet violence continued until 1994 when the horrors of Bosnia burst onto our TV screens. The decade of the '90's, the end of the millennium, would finally shock the world into seeing the heretofore invisible crime of mass rape.

Notes

[1] Niarchos, p. 660.

[2] Donald A. Wells, *War Crimes and Laws of War* (2nd ed, 1991), p. 91, as quoted by Askin, p. 29.

[3] Askin, p. 20, note 48.

[4] Michael Akehurst, *A Modern Introduction to International Law* (1993), p. 12.

[5] Niarchos, p. 660.

[6] Brownmiller, p. 33.

[7] Askin, p. 26.

[8] Hugo Grotius, *De Jure Belli Ac Pacis Libri Tres*, 1646, as quoted by Askin, p. 29.

[9] Niarchos, p. 660

[10] Roy Porter, "Rape — Does It Have a Historical Meaning?", p. 217, in *Rape: An Historical and Social Enquiry*, Sylvana Tomaselli and Roy Porter, Editors (1986), as cited by Askin, p. 21.

[11] Brownmiller, p. 38.

[12] Brownmiller, p. 38.

[13] Askin, p. 28 (Emphasis added)

[14] See Akehurst, pp. 12–13.

[15] Aryeh Neier, *War Crimes* (1999), pp.13–14.

[16] Niarchos, p. 672.

[17] Brownmiller, p. 41.

[18] Askin, p. 41.

[19] J.H. Morgan, *German Atrocities: An Official Investigation* (New York: Dutton, 1916), pp. 81–83, as quoted by Brownmiller, p. 42.

[20] Brownmiller, p. 44. The author cites Harold Lasswell, an early expert on mass communications and propaganda: "A handy rule for arousing hate is, if at first they do not enrage, use an atrocity…These stories (about rape) yield a crop of indignation against the fiendish perpetrators…and satisfy certain powerful, hidden impulses. A young woman, ravished by the enemy, yields a certain satisfaction to a host of vicarious ravishers on the other side of the border." Those few words reveal volumes about how cynically those in power have used sexual assault to promote their own agendas.

[21] Askin, p. 45.

[22] Brownmiller, p. 49.

[23] In the months before the war began, "reports of mob rape directed against Jewish women made their first appearance during the secretly ordered 'spontaneous' riots of November, 1938, the Kristallnacht that began in Munich and spread throughout Germany...The Kristallnacht became the model for a pattern that was to be repeated in many towns in many places once the official war began". Brownmiller, p. 49.

[24] Sheldon Glueck, *War Criminals, Their Prosecution and Punishment* (1996), p. 57, as cited by Askin, p. 55.

[25] Ilya Ehrenburg and Vasily Grossman, *The Black Book* (1980), p. 302, as cited by Askin, p. 55.

[26] Askin, p. 57.

[27] Askin, p. 59.

[28] Askin, p. 60.

[29] Alfred M. DeZayas, *The Wehrmacht War Crimes Bureau, 1939–1945* (1989), p. 179, as cited by Askin, p. 61.

[30] The only mention of rape as a war crime came in the so-called Control Council Law No. 10, the document providing for trial of those individuals who had lesser responsibility during the war and, therefore, were not under the jurisdiction of the Tribunal. Article 2 lists a series of atrocities including rape which constitute crimes against humanity; that is, inhumane acts carried out against a civilian population in a systematic way.

[31] Askin, pp. 97–98.

[32] Iris Chang, *The Rape of Nanking* (1997), p. 91. Chang's account of the Japanese invasion of Nanking — documented from Japanese, Chinese and Western observer points of view — is an excellent analysis of why and how this wartime behaviour occurred. "How...do we explain the raw brutality carried out day after day in the city of Nanking? ... Looking back on the millennia of history, it appears clear that no race or culture has a monopoly on wartime cruelty. The veneer of civilization seems to be exceedingly thin— one that can be easily stripped away, especially by the excesses of war". Yet there were particular historical circumstances which led to the Japanese soldier being "hardened for the task of murdering Chinese combatants and noncombatants alike. Indeed, various games and exercises were set up by

the Japanese military to numb its men to the human instinct against killing people who are not attacking" (p. 55).

[33]Chang, pp. 94–95.

[34]Leon Friedman, ed, *The Law of War, A Documentary History* (1972), Vol. 1, p. 1061.

[35]Lawrence Taylor, *A Trial of Generals, Homma, Yamashita, MacArthur* (1981), pp. 124–125, as cited by Askin, p. 66.

[36]George Hicks, *The Comfort Women*, p. xv.

[37]Askin, p. 203.

[38]Askin, p. 193.

[39]Richard L. Lael, *The Yamashita Precedent, War Crimes and Command Responsibility* (1982), pp. 83–84, as cited by Askin, p. 196.

[40]Askin, p. 201.

[41]Askin, p. 203.

[42]Division for the Advancement of Women, United Nations, *Women 2000*, "Sexual Violence and Armed Conflict: The United Nations Response", p. 3.

[43]Niarchos, p. 679.

VIOLENCE ON THREE CONTINENTS

"As this century comes to an end there is a paradox. Humanitarian law and international human rights have never before been more developed; yet never before have human rights been violated more frequently."[1]

Richard Goldstone, *Crimes of War*

The last decade of the millennium saw more women physically violated than ever before during a period that was ostensibly peace-time. In the short span of just three years, 1991 to 1994, many thousands of women were brutally raped during armed conflicts — more than has occurred at any time other than World War II.

In this chapter, we look at what happened during those three years, in three different countries, on three continents — Haiti in the Americas, Bosnia-Herzegovina in Europe, Rwanda in Africa — to try to draw out from these examples the larger lessons of what conditions spawn and then encourage sexual violence during times of conflict. In one, Haiti, there was no aggressor, only a dictatorial state. In the second, the former Yugoslavia, newly-declared nations fought over territory each claimed. In the third, Rwanda, two ethnic groups shared a country, with one determined to exterminate the other.

Haiti: No punishment, no progress

Haiti is the story of chronic violations of human rights for decade

after decade. It is a story about what happens to a population where
military and other officials are rarely punished and, therefore, have
no fear of continuing their abuses. It is also a story that has not
captured headlines, where it continues to be difficult to gather
accurate information on the violence pervading everyday life.

This small nation lies about 660 miles southeast of the United
States, in the Atlantic Ocean, with the Caribbean Sea to its south
and west. It occupies the western one-third of the island of Hispaniola
which it shares with the Dominican Republic. Nearly 7 million people
live in rough and mountainous terrain, with few natural resources
and a shortage of farmable land. It is the poorest country in the
Americas; its gross national product is among the lowest in the
world.

Most Haitians descend from the half million African slaves who
were set free when Haiti gained its independence from French
colonial rule in the early nineteenth century. Haiti's history has been
marked by unstable governments, with frequent coups and assassinations. Most regimes have protected the upper-class mulatto elite
and neglected the bulk of the population who struggle to survive
by subsistence farming.

For nearly 30 years, from 1957 to 1986, Haiti was virtually the
personal property of the Duvalier family. Under the dictatorship of
first Francois (Papa Doc) and then his son Jean Claude (Baby Doc),
the country suffered extreme repression. Virtually all the institutions
of society were controlled by the Duvaliers: the army, the police,
the courts, the prison system, the trade unions, the press. An
estimated 20 to 30,000 Haitians died on the orders of the Duvaliers
and many, many more went into exile outside of Haiti in order
to stay alive.

In late 1985, the army's killing of four secondary school students
who were part of a demonstration finally triggered an overthrow
of the government. Duvalier and his family fled to a safe, comfortable exile in France and successfully evaded any attempts at
prosecution. A National Governing Council (CNG) was formed to

act as a transition government, overseeing the drafting of a new constitution and presidential elections. But, as ordinary Haitians saw little or no effort to punish anyone from the Duvalier era, they engaged in bouts of vigilante violence.

For the next four years, Haiti barely rose above the level of chaos, as the government passed from the CNG to a series of generals whose tactics continued to be all too similar to those of the Duvaliers. Finally in 1990 a member of the Supreme Court took over as transition president; the country's first free elections were scheduled for December of that year. Late in the campaign, Father Jean-Bertrand Aristede, an activist priest, entered the campaign and electrified it, securing two-thirds of the vote in a heavy turnout.

Aristede took office in February, 1991, and began serious efforts at reform. But within eight months, a military coup had forced him to flee. Aristede had made mistakes including appearing to sometimes endorse vigilante violence. However "once the coup was launched, the army's atrocities quickly dwarfed Aristede's worst failings".[2] The military authority suspended virtually all constitutionally guaranteed rights and procedures

Civilians fair game for military

For the next 3 years, anyone suspected of supporting Aristede was fair game for the military and police; their armed civilian auxiliary, known as attaches; and quasi-political bands of thugs known as *zenglendos*. Poor neighbourhoods were assumed to be loyal to Aristede and were ruthlessly attacked.

In the wake of this violence, an estimated 100,000 Haitians fled the country. Three times that number — most of them men — were driven into hiding within Haiti. This forced displacement of a large part of the population served the aims of the military by undermining both social and political organization.

Women were targeted specifically for sexual assault. These assaults took several forms.

- The military and its allies frequently attacked women in their homes, ostensibly in searches to locate their husbands or fathers, who were accused of being political activists. The violence sent a clear message that a man's political views and activities would be punished by rape of the women closest to him. "...the scenario is always substantially the same. Armed men, often military or FRAPH members, burst into the house of a political activist they seek to capture. When he is not there and the family cannot say where he is, the intruders attack (rape) his wife, sister, daughter or cousin."[3] These assaults were often coupled with destruction of the house and personal possessions.

- Women were also targeted for their own political beliefs, as seen in comments from a 26-year old student who had been an active supporter of Aristede. She was stopped by two men as she walked home at about 7 pm one evening after a visit at a friend's house.

"They asked me my name, where I lived, and what my political opinions were...Then one said, 'I am going to rape you. Tell your boyfriend and your 'Father' (a reference to Aristede) that I am going to rape you'."

Both men did rape the young woman, known only as FF, and then told her to walk "normal like nothing has happened to you". Unlike most Haitian women, FF sought medical attention. But she did not report the assault to the police, saying, "I would never speak to the police...You really risk your life going to talk to them because everyone knows they are part of the crime problem."[4]

- Assaults were also carried out by so-called *zenglendo*, the paramilitary force used by the army to terrorize the civilian population. The line between political oppression and street crime blurred with this group.

In one example of *zenglendo* violence, four armed civilians broke into a church during a prayer service, demanded the church's collection plate, kicked and threatened the parishioners, and demanded to know who was the pastor. Eventually the men grabbed

SE, 24, and RA, 17, took them outside the church and assaulted them. According to the pastor, "when they left with the girls, they said, "You women who are watching, we are going to come back to get you to do the same."[5]

- Women's rights activists in Haiti were also targeted for attacks, with threats of rape and beating. One woman told Human Rights Watch that activists often are "beaten in their female parts — primarily their breasts and abdomen". Another talked of an assault in which a woman activist was beaten by a group of soldiers and told, "We'll beat you until you can't have kids, until you can't have kids like yourself (activist)."[6]

During the three years of military rule, there are no accurate figures on just how many women were raped. Human rights observers went into Haiti periodically to try to assess the situation. In January, February and May 1994, for example, nearly 100 politically-motivated rapes were documented by fact-finding missions from the United Nations/Organization of American States International Civilian Mission, the Inter-American Commission on Human Rights, Human Rights Watch, and the National Coalition for Haitian Refugees.

In 1994, a multinational force led by the United States helped reinstate President Aristede to power and two years later, his handpicked successor, Andre Preval, became Haiti's second elected president. In the years since, serious human rights violations have been stopped but little has been done to punish earlier crimes. The Aristede and Preval governments have decried the human rights atrocities carried out by the military, but have not done much beyond that. Only a few lower level military personnel have been prosecuted; a much lauded truth and justice commission issued a report that the government failed to make public; and the criminal justice system remains weak and ineffective.

Today Haiti again teeters on the brink of violence. Political infighting has immobilized the government and prevented the building of strong institutions, including an effective judicial system.

Sadly, most Haitians have lost faith that they will ever see their suffering redressed. "On the streets, the mood is one of despair with the situation and anger at political leaders..."[7] "Haiti's experience illustrates the dangers of ignoring accountability for past violent abuse in the haste to secure a transition to democracy. Each time a supposedly reformist regime took power, Haitians were asked to forget the past, to look forward to a new era."[8]

For Haitian women, bearing the physical and psychological scars of rape, there is no forgetting.

Bosnia: Broadcasting atrocities

In late 1991, shortly after military rule began in Haiti, people around the world saw and heard, in newspaper and TV reports, initial stories of sexual assaults against women in the part of former Yugoslavia known as Bosnia-Herzegovina. Initially, the reaction was a "worldwide shrug, in effect saying that rape is an unavoidable part of the battlefield...initial stories from Bosnia... (were) viewed as unremarkable by citizens in the West (who were confused by the war itself) and discounted by politicians in the West (lest public alarm at atrocities force them into action)."[9]

By the time the United Nations belatedly reacted in February 1993, by chartering an International Tribunal to investigate human rights crimes, a number of facts were clear:

- At least 20,000 rapes had occurred, 90% of them assaults of Bosnian Muslim women by Bosnian Serb men.[10]
- Rape had been used as a stated strategy and weapon by political and military leaders to achieve their overall goal of acquiring the territory of Bosnia, a newly-declared republic with a majority Muslim population. The intent was to strike the community at its most vulnerable, that is, women and children, in order that their horrific experiences would ensure they would never come back to their home villages.

- Young women were specifically targeted for repeated rape, with the goal of impregnating as many as possible, thus forcing them to bear "little Serbs" and further breaking down both individual women and their families.

The roots of the Bosnian conflict go back centuries, to the Middle Ages when the states occupying the Balkan peninsula in southeastern Europe were split along political and religious fault lines. One-half was claimed by the so-called Western Roman Empire, that of the Catholic church and the Latin alphabet; the other by the Byzantine or Eastern Roman Empire, that of the Orthodox Christian church and the Cyrillic alphabet. Croatia and Slovenia became part of the Western world, Serbia part of the Byzantine world.

For a few decades in the first half of the 1300s, the Serbs expanded their empire from Belgrade in the north, to present-day Bulgaria in the east, and central Greece in the south. The Ottoman Turks ended that empire by winning the famous battle of Kosovo, which began 500 years of Muslim Turk rule over Orthodox Christian Serbs. During those centuries of domination, the Serbs' rights were severely curtailed and they developed a deep-seated distrust of the Muslim world.

Modern history has seen these fault lines reinforced. The federation of Yugoslavia was conquered by the Nazis during World War II. In Croatia the Nazis set up a puppet government, modeled after fascist Germany, and installed a Croatian leader from the Ustasha, the far-right group whose goal was to exterminate the Serb population. They succeeded in killing between 60,000 and 70,000 Serbs.

Serbia itself was kept directly under German military rule. A resistance movement formed calling themselves the Chetniks, but they soon ignored their occupiers and instead took revenge on Croats and Muslims for the Ustasha atrocities in Croatia.

As Tito seized power at the end of the war, he continued the cycle of violence by ordering his troops to summarily execute thousands of people he saw as enemies: an estimated 30,000

Croatian and Slovenian soldiers in Bleiburg, another approximately 30,000 Chetniks near the Serbian-Bosnian border, and thousands of civilians as well.[11]

By the time Tito formed a communist federation in 1946, historians believe that more than a million and a half people had died in Yugoslavia, so many that nearly every family had lost some relative to the Ustasha, the Chetniks, or the troops loyal to Tito.

Tito's federation plastered over political, ethnic and social differences with a strict regime that severely punished any show of nationalism. In spite of (perhaps because of) that fact, the differences festered — particularly among rural people who had little exposure to other ethnic groups. In fact, in urban settings, the story could be quite different; for example, Sarajevo was considered a model of multiethnic living.

After Tito's death in 1980, Yugoslavia began to disintegrate. As Slobadan Milosevic rose to power in Serbia, his divide-and-conquer tactics were the opposite of Tito's. He used "all the key institutions in Serbia — the academics, the media, the Serbian Orthodox Church (to) set Yugoslavia's ethnic and national groups against one another".[12] As early as 1986, there are reports of a "secret memo" drafted by the Serbian Academy of Arts and Sciences, calling for the establishment of a "Great Serbia". Some observers read it as a "confused, racist rallying cry" that is the skeleton of a plan for military aggression and even genocide.[13]

Around this same time, the Yugoslav army was slowly purged of non-Serb soldiers. First came discrimination against non-Serbs, including pay cuts and increased workloads, then came an unusually high number of serious accidents during basic training. "The number of dead recruits was finally great enough to make it clear that the military murder of non-Serbs within the 'national' army was systematic and genocidal. Mothers in Zagreb...went to Belgrade (and) demonstrated...(they) were manhandled, beaten, arrested, and sent back to Zagreb."[14]

In 1990 the republics of Slovenia, Croatia and Serbia held their first free parliamentary elections. Reformist Milan Kucan took over in Slovenia.

Right-wing nationalist Franjo Tudjman won in Croatia; his government would later use the old Ustasha symbol on its flag. Ultra nationalist Milosevic took over in Belgrade, rejecting all suggestions of a confederation of the three republics and declaring that Serbia existed wherever Serbs had settled during their history. That meant Croatia, Bosnia, Montenegro, Kosovo, and Vojvodina.

By the spring of 1991, Croatia and Slovenia had decided there was no alternative but to declare themselves autonomous states. Both parliaments did so on June 25. The Serbs went on the attack using the Yugoslav Federal Army which was by that time a Serbian army. Slovenia was well armed and fought off the federal army. One-third of Croatia was conquered by the time the international community brokered a ceasefire of sorts. Between Croatia and Serbia lay Bosnia, Milosevic's next target.

"Ram" plan: Assault the most vulnerable

The Serb strategy for Bosnia and Croatia was laid out in meetings where the Milosevic government's plans were discussed and ratified, and later reported by Italian journalist Giuseppe Zaccaria who saw photographs of the meeting minutes. In the case of secessionist movements, the so-called Ram Plan ("ram" being loom...the plan weaving its way across BH and Croatia) "foresees the (Serb) occupation of Croatian territory and Bosnia, the movement of (Serb) troops into the Sandjak, and the control of an area where Muslim fundamentalists might prove to be particularly strong."[15]

The minutes go on to talk about a variation of the Ram Plan written by the army's special services, including psychologists and experts in psychological warfare. They argue that particular methods should be adopted if a widespread conflict develops:

"Our analysis of the behaviour of the Muslim communities demonstrates that the morale, will and bellicose nature of their groups can be undermined only if we aim our action at the point where the religious and social structure is most fragile. We refer to the women, especially adolescents, and to the children. Decisive intervention on these social figures would spread confusion among the communities, thus causing first of all fear and then panic, leading to a probable retreat from the territories involved in war activity."

"In this case, we must add a wide propaganda campaign to our well-organized, incisive actions so that panic will increase. We have determined that the coordination between decisive interventions and a well-planned information campaign can provoke the spontaneous flight of many communities."[16]

This is late 1991, just as the first reports of rape death camps in Bosnia are reaching Zagreb.

Italian journalist Zaccharia also reports on other documents, including a letter from a Serb battalion commander, written to the chief of the secret police in Belgrade.

"Sixteen hundred and eight Muslim women of ages ranging from 12 to 60 years are now gathered in the centres for displaced persons within our territory. A large number of these are pregnant, especially those ranging in age from 15 to 30 years...the psychological effect is strong and therefore we must continue (the practice of genocidal rape)".[17]

The path was clear and we see it now, with the benefit of hindsight. The Serbs had a defined policy of clearing out territories they wanted, not by conventional warfare, but by attacks on civilians — and not just any civilians, but on women, whose assault would cause the most trauma to their communities. Systematic rape and forced pregnancy were the tools they used.

Why forced pregnancies? On the simplest level, it was one more trauma that helped to tear apart the lives of women and their families. On a more complex level, it fed the Serb fantasy of their bloodline being one of a superior race, with superior genes, so that any child conceived with Serb sperm would be a little Serb.

"One woman was detained by her neighbour (who was a soldier) near her village for six months. She was raped almost daily by three or four soldiers. She was told that she would give birth to a chetnik boy who would kill Muslims when he grew up."[18]

The patterns of sexual assault were visible by the time the Bassouni Commission, a fact-finding team set up by the United Nations, had completed its work in 1994. Most of the rapes were committed between April and November 1992; fully 80% of the assaults took place while the women were in custody; and about 90% of the victims were Muslim women.

The UN report groups the sexual assaults in Bosnia into five categories:

- Rapes committed by small groups, often paramilitaries, before any fighting started. The goal was to terrorize the inhabitants and "encourage" them to flee by assaulting village women, usually in full view of their families or others. In one typical case, a woman was gang raped by 8 soldiers in front of her 6-year old sister and 5-month old daughter.

- Rapes committed as towns and villages are invaded and captured, to assure that villagers would never want to return to their homes. Residents were assembled and prepared for deportation. Women would be selected to be raped by several men in front of their neighbours; for example, one women reported seeing the rape of an elderly women in front of a group of 100 villagers.

- Rapes occurring after women and men were placed in separate detention camps, once the town or village was cleared. "Soldiers, camp guards, paramilitaries and even civilians may be allowed to enter the camp, pick out women, take them away, rape them and then either kill them or return them to the site."[19]

- Rapes committed in about 30 so-called rape death camps, some large and well organized, housed in factories, schools and hospitals; others small and informal, in cafes, homes, even barns. "In those camps, all of the women are raped quite frequently, often in front of other internees, and usually accompanied by beatings and torture."[20] "It's very important to understand the term rape/death camp...it is not a camp for survivors...only 20% of captured women survive...That is the point of ethnic cleansing."[21]

- Rapes committed in forced brothels, in hotels or other facilities where. These women were kept "for the sole purpose of sexually entertaining soldiers... These women are reportedly more often killed than exchanged."[22]

Brutality, humiliation, torture

Wherever the rapes took place, they shared common characteristics. They were brutal, often involving extreme humiliation and sexual torture. Here a 40-year-old woman tells of her experience in one of the rape camps:

"They liked to punish us. They would ask women if they had male relatives in the city; I saw them ask this of one woman and they brought her fourteen-year old son and forced him to rape her.

"If a man couldn't rape (ie, if he was physically unable) he would use a bottle or a gun or he would urinate on me."[23]

Most were gang rapes, and many were made into public spectacles, committed in front of other soldiers, family members, neighbours, other detainees, no doubt to increase the humiliation of the victims and the terror of viewers.

On women described her experience:

"These soldiers would invite their friends to come watch the rapes. That was like in the movie theatre. All sit around while others do their job...Sometimes those who were watching put out cigarette butts on the bodies of the women being raped."[24]

Sometimes the rapes really were theatre: the rapes were video-taped and shown on Serbian television but the victims were pre-sented as Serbian women and the attackers as Muslim or Croatian men.

Often those who raped Bosnian women were friends, neighbours, acquaintances from nearby villages. Survivors report that some initially refused and were severely punished, even killed. Others were incapable of raping on demand, and victims tell of their attackers using pornography, drugs or alcohol as stimulation.

One rapist told his victim:

"We have to do it, because our commanders ordered it, and because you are Muslim — and there are too many of you Muslims. We have to destroy and exterminate you, so that the heroic Serbian people can take over the reins in this area again."[25]

There could hardly be a clearer statement of ethnic cleansing, that is, "using violence and deportations to remove any trace of the other ethnic communities who had previously cohabited with Serbs in the coveted territories. This 'cleansing' was the goal of the war, not the unintended consequence...Simply put, achieving ethnically homogeneous States in a region of historic mixing could not be achieved except through extreme violence."[26]

In fact, the widespread rape of women may, in fact, be seen as yet another form of genocide, especially in a traditional patriarchal society such as Bosnia. "The mass rape of women has as its purpose the ruin of women as future wives and mothers, or the wrecking of their marriages. In Islamic society modesty is highly prized; by tradition, many women still will not allow themselves to be seen naked even by their husbands. One can imagine the attitude towards a woman who has been raped by many men, daily, for months. And her despair..."[27]

Despite early evidence of genocidal practices first in Kosovo, then in Bosnia, the international community did not intervene for over three years. When they did finally establish the United Nations International Tribunal in February 1993, they did not commit the resources needed for the Tribunal to do its work properly. No chief prosecutor was even named until July 1994 and it was only a year later that the Tribunal first indicted two major figures in the war.

Today, the Tribunal continues its painfully slow work. Only in 1996 did it bring the first charges against those responsible for rape and sexual torture.

Rwanda: 100 days of atrocities

On April 6, 1994, a plane was shot down as it landed in Kigali, capital of Rwanda. Among the dead were the presidents of two nations, Rwanda and Burundi, who were returning from Tanzania where they were involved in negotiations to end a three-year war in Rwanda. The plane drew fire near the camp of specially-trained and recruited Presidential Guard, who most likely committed this act of sabotage.

The crash triggered a countrywide nightmare that left an estimated 1 million dead, 3 million refugees, and between 200,000 and 500,000 women the victims of sexual assault.

The conditions for this nightmare had been building for many years. The minority Tutsis had ruled the country for two centuries before Rwanda became a colony of first Germany, then Belgium. Under Belgian rule ethnic differences were accentuated: the cattle-rearing Tutsis were given higher social status and privileges, such as access to western education, that were denied the majority Hutus who were primarily peasant farmers.

In the early 1960's when Rwanda gained independence, the Hutus overthrew the Tutsi monarchy and established a republic, under Hutu leadership. But by 1990 the government was at war with an invading force consisted largely of Tutsi refugees who fled Rwanda during the revolution. Bitter fighting had resulted in about 10% of the population being displaced, some 700,000 people. The invasion was seem by many Rwandans as an attempt by the Tutsi to reestablish the monarchy.

In addition, President Juvenal Habyarimana's regime, in power for two decades, was being pushed by forces from both inside and outside the country, to allow more freedom, broaden political representation and permit other political parties. A powerful elite within the ruling regime was increasingly uneasy with these changes. That powerful elite had, in fact, been planning what can only be called genocide against the Tutsis for months. The president's death was simply the trigger. In fact, the commander of a United Nations peacekeeping force in Rwanda had warned in January of a Hutu plan to exterminate Tutsi; he was ordered to take no action.

For several months before violence began, the state-controlled radio station had broadcast racist messages, masked with the glib language of program hosts who used "street slang, obscene jokes and good music...people went on listening to it with a kind of stupefied fascination, incredulous at the relaxed joking way in which it defied the most deeply cherished human values".[28] Obscene cartoons lampooning Tutsi women also appeared, and extremist propaganda specifically attacked the sexuality of Tutsi women as a tool by which Tutsis had tried to control Hutu men.

Once the initial fighting began in the capital, radio stations controlled by the Presidential Guard began to broadcast intense hate messages against the Tutsis.

It encouraged, even directed both killing and sexual violence as ways of achieving the political goal of dehumanizing, subjugating, and destroying the Tutsis as a political group. Targets also included Hutu women who were affiliated with the political opposition, married to Tutsis, or had attempted to protect Tutsis. Young girls and those considered especially beautiful were at the mercy of the militia who, regardless of politics, raped indiscriminately.

Leaders in the capital ordered local officials to incite their communities as well; those who refused were quickly replaced with those who would.

In the 100 days that followed, somewhere between 500,000 and one million people, mostly Tutsis, were slaughtered, many of them by fellow villagers. Some observers believe rural Rwanda was ripe for incitement. Arms were prevalent because of the ongoing war. And, there were large numbers of young men with "no land, no jobs and no chance of ever gaining access to either".[29]

Even so, those factors would not likely have been sufficient in a less rigid society. In fact, there had always been a strong tradition of unquestioning authority in the pre-colonial kingdom of Rwanda. This tradition was of course reinforced by both the German and Belgian colonial administrations. And since independence the country had lived under a well-organized tightly-controlled state. "When the highest authorities in that state told you to do something you did it, even if it included killing."[30]

As many killers as victims

Whatever the reasons, the Rwandan genocide is unique in the number of people incited to kill: perhaps as many killers as there were victims, that is, 800,000 to one million. "For a tiny country with a total population of about 8 million (fewer than half of them adults), the number of probable killers is even more mind-boggling than the number who died...By engaging so large a portion of the population in the slaughter ...Hutu organizers of the genocide made hundreds of thousands of ordinary Hutus bloody their hands. In so doing, they deepened the hatred of the victims and spread the guilt...If all are guilty, no one is guilty.

"The openness with which the slaughter took place indicated confidence that no external force would intervene and that the organizers would be secure against reprisals. On this conclusion, they were on solid ground. The war in Bosnia had been under way for more than two years, and little had been done to stop it."[31]

And, little was done in Rwanda: The international community again hesitated to intervene, fearing they would be bogged down in an ill-understood and never-ending conflict. Governments were particularly reluctant to step into an African conflict, because of a recent debacle in Somalia where a United Nations force suffered casualties and had to withdraw from the country without having pacified it.

So the 100 days of violence continued. No one knows just how many women were raped in Rwanda since most of them were murdered afterwards. Estimates range as high as 200,000 to 500,000 based on the number of pregnancies which resulted from the rapes. The Rwanda National Population Office estimates between 2000 and 5000 pregnancies resulted from the rape of women in 1994. It is generally accepted that unprotected intercourse will result in pregnancy 1% to 4% of the time, thus the estimates of several hundred thousand rapes.

Whatever the exact number, the impact was horrific, particularly given the short time period. Since most of the assaults and murders took place in a 100-day period that means some 5000 murders and 2,000 rapes were committed *each day*. The sexual assaults "were frequently part of a pattern in which Tutsi women were raped after they had witnessed the torture and killings of their relatives and the destruction and looting of their homes. According to witnesses, many women were killed immediately after being raped".[32]

Alive but dying

Those allowed to live were frequently left pregnant and diseased. Many were intentionally raped by men who knew they had AIDS, with attackers sometimes actually telling their victims that they wanted them and their children to live to infect other Tutsis – especially the returning male refugees. AIDS is about twice as prevalent in Tutsi women who were raped during the genocide as in the general population, where one in four people are infected.

Immacule, a 32-year old survivor interviewed by the BBC, told of being raped by 8 men after her entire extended family of 50, including her husband and three children, were massacred while hiding in their church. She alone survived as she fell beneath their bodies.

She suffered from excruciating abdominal pain for nearly 5 years, afraid that she would be criticized and isolated in her community if she admitted to being raped and sought treatment for the after-effects. To have survived is seen as shameful by some Tutsis who fled the country, says Immacule. "They think you gave your body to survive."

When Immacule began to develop ulcers and abscesses, she overcame her shame and fear, and visited a women's clinic for treatment. There she learned she needed surgery for a serious pelvic inflammation — and that she had contracted AIDS. "I don't even have anyone to bury me," she says bitterly. "I wish I had the courage to kill myself. I am alive but this is a living death."

The same is true for Biatta, who was held as a sex slave by her neighbour after he had murdered her husband. When the massacre ended, and her neighbour fled, she found herself pregnant but with no money for an abortion. She delivered a daughter, now four years old, whom she dutifully cares for but finds difficult to love. As a final legacy of her rapist, both Biatta and her daughter have AIDS. Biatta has not yet had the courage to tell her three older children that they will soon be orphans.[33]

Other survivors told their stories to the UN Special Rapporteur on violence against women, Radhika Coomaraswamy, when she visited Rwanda in November 1997.

- Monique and her 6-year old daughter ran into the forest to hide when the killing started. She was then six months pregnant. The militia found them and raped both her and her daughter repeatedly. Monique's uterus was damaged and she subsequently delivered an abnormal baby. Her young daughter still bears the physical and psychological scars of her rapes.
- A journalist came with a group of men to Donatilla's house to rape her. "Two of the men kept her legs apart while the journalist, using rusty scissors, cut her genitalia....Her aggressor then displayed the cut genitalia in public for everyone to see."[34]
- Jeanne was raped by one of her neighbours at the entrance to her church. "Her neighbour told her: 'I have AIDS and I want to give it to you.' He then raped her, right in front of the church, even though she was pregnant. The other two men also raped

her afterwards. Jeanne survived the genocide, but now she has AIDS and she is wrecked with pain."[35]

Health problems resulting from sexual assaults and resulting pregnancies are a major hurdle for Rwandan women as they struggle to rebuild their lives. Some, like Jeanne, Immacule and Biatta face a bleak future as they begin to see AIDS ravage their bodies. Some have not even sought medical care, in spite of chronic pain, because of their deep sense of privacy and shame about what has happened to them. Others have no access to adequate medical care; there are only five gyneacologists in all of Rwanda. And, almost all bear the trauma of seeing those they loved slaughtered, then being themselves assaulted and humiliated, with all the accompanying feelings of depression, anger, and hatred.

Poverty is another major problem for Rwandan women, especially since so many of the male relatives on whom they depended were killed during the genocide. Their situation is exacerbated by their second-class status under Rwandan law: women cannot inherit property unless specifically designated as beneficiaries. As a result, thousands of women cannot lay legal claim to their family's home, land or bank account.

The UN International Tribunal for Rwanda, even more than its counterpart in the former Yugoslavia, has suffered from inadequate funding, staff resources, and political support. Only in 1998, four years after its founding, did the Tribunal issue its first rulings which did include charges of rape and sexual abuse as both a crime of genocide and a crime against humanity.

But, the sheer magnitude of the genocide and the number of perpetrators mean that justice in Rwanda will be slow and limited. Today over 100,000 Rwandans remain in prison, waiting to be tried on charges of murder and other crimes associated with the genocide.

Sadly, the conflict goes on even as Rwandans start the long process of healing their society. Ethnic fighting continues in the northwest of the country, spilling over into the eastern part of the Congo. And the women of Rwanda struggle to live again as normal people in a country so full of bloodshed and atrocities that few are guiltless.

Notes

[1]Richard Goldstone, *Crimes of War*, p. 16.

[2]*Rape in Haiti: A Weapon of Terror*, Human Rights Watch/National Coalition for Haitian Refugees, p. 6.

[3]*Rape in Haiti*, p. 7.

[4]*Rape in Haiti*, p. 12.

[5]*Rape in Haiti*, p. 15.

[6]*Rape in Haiti*, p. 7.

[7]Larry Rohter, "Political Feuds Rack Haiti", *New York Times*, Oct. 18, 1998.

[8]*Thirst for Justice: A Decade of Impunity in Haiti*, Human Rights Watch, 1996, p. 4.

[9]Thom Shankar, *Crimes of War*, p. 323.

[10]By far the greatest number of rape victims were Bosnian Muslim women. But, it is important to note that Croat, Kosovan and Serbian women were also brutally assaulted, as the violence spread among all groups of combatants. Even men were not safe: there are a number of reports of terrible sexual assaults on Bosnian Muslim men as well.

[11]Stiglmayer, *Mass Rape*, pp. 11–12.

[12]Florence Hartman, *Crimes of War*, p. 51.

[13]Allen, p. 49.

[14]Allen, p. 51.

[15]Allen, p 56.

[16]Allen, p. 57.

[17]Allen, pp. 59–60.

[18]Askin, p. 272.

[19]Niarchos, p. 655.

[20]"Final Report of the Commission of Experts Pursuant to Security Council Resolution 780" (UN Doc s/114/674, Annex), paragraph 247.

[21]Askin, p. 277.

[22]"Final Report" as cited above, paragraph 249

[23]*War Crimes in Bosnia-Herzegovina*, Human Rights Watch (Vol. 2, 1993), p. 216, as cited by Askin, p. 279.

[24]Catherine MacKinnon, "Turning Rape into Pornography", in *Mass Rape*, p. 78.

[25]*Die Welt*, October 1, 1992, as quoted by Askin, p. 272.

[26]Florence Hartman, *Crimes of War*, p. 52.

[27]Askin, p. 270.

[28]Gerard Prunier, *The Rwanda Crisis: History of a Genocide* (New York: Columbia University Press, 1995), p. 224, as cited by Neier, p. 207.

[29]David and Catherine Newbury, "Rwanda: Politics of Turmoil", Washington Office on Africa, April 1994, p. 2.

[30]Gerald Prunier, as quoted by PBS and WGBH Frontline, 1998.

[31]Neier, pp. 217, 227.

[32]"Shattered Lives: Sexual Violence during the Rwandan Genocide and its Aftermath", Human Rights Watch, September 1996, p. 1.

[33]Both stories were told as part of a BBC TV documentary, "Condemned to Live", broadcast on BBC World Forum in October 1999.

[34]As above, p. 8.

[35]As above, p. 8.

ASSAULTS IN ASIA

"Long after slavery was abolished in most of the world, many societies still treat women like chattel: Their shackles are poor education, economic dependence, limited political power, limited access to fertility control, harsh social conventions and inequality in the eyes of the law. Violence is a key instrument used to keep these shackles on."[1]

Charlotte Bunch, UNICEF 1997 Annual Report

Asia is an immense continent of unequalled diversity. Its 20 largest countries account for more than 3 billion people, over half of the world's total population. China and India — the world's most populous countries and home to two of its most ancient civilizations — today are developing nations struggling to build strong, modern economies that can sustain their immense and ethnically diverse populations.

Indonesia, with the fourth largest population in the world at 212 million, encompasses nearly 1,000 populated islands and a total land mass 20 percent larger than the United States. Its people represent more than 300 ethnic groups, speaking more than 200 indigenous languages.

Meanwhile, homogeneous Japan — tiny by comparison in both land mass and its population of 125 million — boasts the world's second largest economy, following the United States.

Throughout this diverse landmass, women have worked for decades to improve their position in society. The 1990's were no exception. Many Asian women struggled to maintain their physical safety and integrity as they faced a variety of obstacles: in some cases, weak political institutions, punitive laws or dictatorial governments; in others, ethnic and religious conflicts, civil wars, or outside invasions.

Many endured rape and other forms of sexual abuse, in patterns of violence familiar from earlier chapters.

- Women were assaulted based on their ethnicity. In Sri Lanka, for example, a 15-year war between the government and militants from the minority Tamils, continues to generate major human rights abuses. In 1998 the UN Special Rapporteur on Violence Against Women reported a number of incidents of rape, mutilation and murder carried out by the military against women and girls of Tamil origin in the north and east of the country. In general, soldiers and police committing these crimes are rarely punished or held accountable for their actions.[2]
- Women were tortured and raped because warring factions accuse them of supporting "the enemy". In Nepal, torture and rape have been widespread since February 1996 when the Communist Party of Nepal declared a people's war against the government with abuses committed by both parties to the conflict. As is typical, each side targets those civilians viewed as supporting the enemy, with female detainees raped and sexually humiliated.[3]
- Women were attacked based on their husband's political affiliation. During three decades of war and political upheaval in Cambodia, rape and sexual harassment of women were committed by all factions in the continuing conflict. Women were often targeted for rape based on their husband's alleged political affiliation, or on their race; for example, Vietnamese living in Cambodia were particular targets. Women have been forced to migrate to other parts of the country or to flee to refugee camps,

usually in Thailand, where they are still subject to sexual attacks by soldiers and security officers.[4]

• Women were subjected to violence by their country's soldiers, who view sexual assault as a reward or right. In Myanmar the all-powerful military has been free to rape and sexually harass women in their own villages, as they cross military checkpoints, or when they serve as forced labourers in military camps.

• Women were raped, beaten, even murdered by military or paramilitary groups in order to humiliate and terrorize a local population. In India, in the state of Kashmir, gang rape of many, even most women, in a village has been a strategy for frightening a population so they would not support insurgent groups.

• Women and girls were taken by the military to become sex slaves, or to be sold into prostitution. In East Timor, Indonesia, the practice of soldiers forcing village women to become "local wives" was widespread, particularly among poor, uneducated women who were too frightened to resist.

In this and the next chapter, we look in detail at four countries in Asia — Afghanistan, Myanmar, India and Indonesia — where human rights groups have been able to fully document the ways in which sexual violence has been used against women and where, sadly, much of the violence continues.

Afghanistan: Resistance groups versus civilians

This rugged mountain nation has been an important crossroads of Central Asia for hundreds of years, sitting astride historic trade and invasion routes between China and South/Southeast Asia. Afghanistan's mix of cultures and languages reflects its location: the Pustun, Tajik, Hazara and Uzbet are the four largest ethnic groups, with at least half a dozen smaller groups. Afghan Persian or Dari is spoken by most of the people, at least as a second language, with each ethnic group speaking its own language or dialect as well.

In the mid-18th century a prince from the majority Pashtun group united the country's multiple chieftainships, principalities and provinces into a single kingdom, one which ruled for more than two centuries. In the 19th century Afghanistan, like numerous Asian countries, became a battleground for competing colonial powers, in this case Britain and Russia. The British gained control after waging two bloody wars against Russia, then insured Afghan cooperation by supporting the king's rule and contributing heavily to royal coffers.

Afghanistan gained its independence from Britain in 1919, and the royal government began the process of modernization. Even then, there was debate about the character of this modernization between more liberal, Western-oriented city dwellers and a more conservative religious population in the rural areas.

By the 1950's the Afghans had sought military arms and training from the Soviets, after having been rebuffed by the United States. Over the next two decades the Soviets trained all top Afghan military men, and in doing so gained considerable influence in Afghanistan. By the 1970's the Soviets had expanded their role in Afghan politics: first, they helped to engineer a military coup which toppled the king; then they helped to overthrow his successor and install a Communist led "Revolutionary Council". The latter oppressed the Muslim majority and generated enough unrest to bring on civil war.

As the government weakened, the Soviets invaded Afghanistan. For the next 9 years, they occupied the country and battled resistance groups. In the process thousands of civilians and over 15,000 Soviet soldiers lost their lives. Finally, the Soviets withdrew their troops in 1989.

However, the years that followed saw continued conflict, as the Soviet-backed government struggled to hold power against competing resistance forces known as Mujahideen. In 1992, the Soviet-backed government of President Najibullah was overthrown by a combination of resistance groups and some army allies. The Mujahideen then began fighting each other to win control of Kabul and other key cities. The situation deteriorated rapidly as warlords tried to seize greater power for themselves, took control of various territories, and inflicted abuse on the local civilian population.

By 1995 the Taleban, a popular religious force drawing members largely from the majority Pushtu ethnic group, had gained the upper hand and by the following year they displaced the ruling members of the Afghan government. Today the fundamentalist Taleban control the capital of Kabul and approximately two-thirds of the country including the predominantly Pushtu areas of southern Afghanistan. They have wiped out most of their rivals with the exception of the leaders of the largely ethnic Tajeks in the north of the country.

As a result of nearly 20 years of conflict, Afghanistan is one of the poorest and least developed countries in the world: life expectancy is a low 44 years versus 61 years for all of South Asia; maternal mortality is more than 4 times that of India; and a third of all children are destined to die before the age of 5. About two-thirds of Afghans are illiterate, with only 4% of girls and 27% of boys attending primary school.

Rape to intimidate, punish and reward

Throughout the 1990's, women have been particular victims in the continuing conflict.

In the first half of the decade, armed groups — both Mujahideen and warlords — brutally beat and raped women. Many women were abducted, raped and then taken as "wives" by commanders or sold into prostitution. "As territory changes hands after long battles, an entire local population can be subjected to violent, retaliatory punishments. The conquerors often celebrate by killing and raping women and looting property."[5]

Between April 1992, when Mujahideen groups took power in Kabul, and late 1994, Amnesty International documented a multitude of abuses of the civilian population, based on in-depth interviews with refugees.

Rape appeared to be condoned by factional leaders as a method for intimidating conquered populations and for rewarding soldiers.[6]

- A 15-year old girl reported that armed men repeatedly raped her and shot her father for allowing her to go to school. "It was nine o'clock at night. They came to our house and told him they had orders to kill him because he allowed me to go to school...They killed my father right in front of me...I cannot describe what they did to me after killing my father."

- A young woman who had fled Kabul after heavy fighting there told of her father's experience after she left. "One day when my father was walking past a building complex, he heard screams of women coming from an apartment block which had just been captured by forces of General Dostum. He was told by the people that Dostum's guards had entered the block and were looting the property and raping the women."

- Several refugees told the story of a young woman who had lived in Kabul in early 1994. "Her husband had been killed in a bomb attack. She had three children of between two and nine years old. One day she leaves her children to go and find some food. Two Mujahideen armed guards arrest her in the street and take her to their base in a house where 22 men rape her for three days. She is then allowed to go. When she reaches her home she finds her three children have died of hypothermia. She has now lost her sanity and lives in Peshawar."

Other women were targeted because they were from ethnic minorities considered by soldiers to be their enemies.

- "...the forces of General Dostum came to the city...these guards were only looking for Pastun people. We were not Pashtun, so at least our lives were spared...They carried out a lot of atrocities. For example, a number of young women in our street were raped by them. One young women was taken away...a few days later her body was found somewhere in the city."

Sex slaves and prostitutes

Still other women, young girls and even young boys were abducted to become sex slaves of commanders, or to be sold into prostitution.

- "We were a farming family. There were 10 of us...One Jamiat-e Islami commander who had three wives came with his armed guards to our house asking to marry my sister who was 15 years old. My brother objected... But the commander's guards beat my brother...we were forced to give my sister away."
- Another commander, this time of Hezb-e Islama, wanted a young schoolgirl named Farida. The commander contacted the father and asked him to give his daughter to the commander. The family rejected this. The commander then came back in the evening with a number of armed guards and took the girl away.
- A woman in Kabul tells of seeing armed guards taking away young girls and boys. "I was baking bread in my flat one day when I saw armed guards entering the building complex. Some went to the basement of an apartment block and came out carrying young boys aged 12 and 13. They were screaming but the guards forced them to go with them."

Thousands of women fled their homes, only to be assaulted during their trip or in the refugee camps on the borders of Afghanistan and Pakistan. Most camps are controlled by one or another of the armed factions. Women, particularly those not accompanied by men, are assaulted or forced to exchange sex to get access to vital rations.

Given the extent to which the honour of women in Afghan society is seen as all-important, several Afghan women have been documented as committing suicide rather than submit to rape.

- "Nadia was a 16-year old high school student... In mid-1992 her house was raided by armed Mujahideen guards who had come to take her. The father and family resisted. Nadia ran to the fifth floor of the apartment block and threw herself off the balcony. She died instantly. Her father put her body on the bed frame and wanted to carry it in the streets to show the people what had happened to her, but the Mujahideen groups stopped him."

In recent years, women have continued to suffer these kinds of attacks in provinces where conflict continues among the Mujahideen groups and the Taleban. In the two-thirds of the country where the Taleban rule, women suffer a different kind of violence, based on the strict laws governing them. "The rigorously enforced discriminatory edicts include prohibitions against women working, leaving their homes without being escorted by a close male relative, receiving medical treatment from a medical doctor or in a non-segregated hospital, and appearing in public without being fully covered."[7]

If a women is not completely covered by the burda when she ventures outdoors, if even a small amount of ankle shows between the burda and a shoe, she risks having her legs beaten. No make-up is allowed; there are reports of women's lips being mutilated if they are suspected of wearing lipstick. Girls are not allowed to attend school, or women to leave home to work, at least in Kabul. In outlying areas, some women are permitted to work in specific occupations: for example, as health workers serving women only.

Thus, while specifically sexual violence is no longer widespread, Afghan women remain extremely vulnerable subject to the dictates of a repressive and violent regime.

Myanmar: Repressive military regime

Myanmar, formerly known as Burma, lies in Southeast Asia, sharing borders with Thailand, Laos and China on the east and northeast, India and Bangladesh on the west. It is home to about 48 million people in 8 major ethnic groups: Ethnic Burmans make up about 70% of the population, with Shan, Karen, Mon, and Karenni the largest minority groups. Almost 90% of the population are Buddhists.

Primarily an agricultural country, Myanmar was once one of the most economically promising nations in Asia, with oil, gas and mineral reserves, tropical hardwood forests, and a surplus of rice which was exported. Sadly, years of military rule, economic mismanagement, and corruption have made Myanmar one of the poorest countries in the world today. Forty percent of the national budget goes to military spending, only 5% to education, and the average annual per capita income is estimated to be less than US$500. Myanmar is now the world's largest producer of illicit opium and heroin.

A British colony during the 19th and early 20th century, Myanmar gained its independence in 1948. The resistance movement which fought against the Japanese in World War II developed into the main political force during the post-war period. Its leader was General Aung San who was broadly revered among her fellow countrymen as the leader of the independence struggle, and thus able to unify the country by winning the trust of both majority Burmans and ethnic minorities.

The assassination of General Aung San in 1947, just as independence was in sight, left a tremendous vacuum and gradually led to distrust among the various ethnic groups and revolts by communist and other insurgent groups. In 1962 the military made the shaky government an excuse to step in and institute a system in which the military has been supreme in virtually every aspect of life.

As the economy worsened under a closed, authoritarian government, there were occasional outbursts of social unrest which were quickly put down by the military. In 1987, when a poor harvest led to severe rice shortages, the government announced that it intended to tackle the economic situation by withdrawing certain denominations of banknotes from circulation, with no compensation. The already poor population was desperate, and took to the streets in riots and demonstrations which were violently suppressed by the police.

Finally in August of 1988, tens of thousands of protesters went to the streets, first in the capital of Rangoon, then in other cities.

They demanded the government's resignation, an end to the centrally-run economic system, and a new democratic form of government. The response was swift and brutal: thousands of unarmed civilians were killed and a new, even more repressive leadership body established: the State Law and Order Restoration Council (SLORC).

However, Aung Sang Suu Kyi, daughter of General Aung San, and the leader of the National League for Democracy movement, continued to campaign for democratic rule. She was placed under house arrest but still led her party to an overwhelming victory in the 1990 elections which the government allowed to take place. But, the military refused to honour the election results and hand over power; instead they arrested opposition leaders.

In the last decade, the situation in Myanmar continued to deteriorate despite the fact that the government began to open the economy somewhat, and increased economic activity fostered an appearance of greater normalcy. In fact, violence against civilians continued to be a fundamental part of the government's military strategy to wipe out any resistance. The approach is two-pronged. Troops secure resources from local population, especially food, combatants and workers. Soldiers also weaken resource base of insurgent groups by systematically destroying villages and moving the rural population so as to disrupt agricultural production.

Minorities as enemies

These military actions have led thousands to flee into often primitive refugee camps in Thailand and Bangladesh. In addition, the United Nations estimates that half a million people from the minority states of Mon, Karen, Shan and Karenni have become internal refugees, as the military forces the relocation of those suspected of harbouring insurgents. After years of conflict between these minority states and the central government, many soldiers treat the entire minority population as if it were an enemy.

One of the most comprehensive and damaging relocations has taken place in Shan State, the largest of the seven minority states where about 4 million Shan live together with an equal number of Burmans and other minorities. By 1998, over 300,000 people had been moved from 1400 villages. As a result, some 80,000 Shan fled to Thailand. (The Shan are ethnically related to the Thai, have a similar language, and live in southern China and northern Thailand as well as in Myanmar.)

Because families are desperate for food and work once relocated, individuals often risk returning to their villages to see what they can salvage. They often meet with violence.[8]

- 30-year old Nan Ing from Wa Na San village returned to her village to retrieve their rice after relocation. She was caught by three soldiers who claimed she had given rice to insurgents. They first raped her, then poured boiling water over her body, so that she was burned from her neck to her feet. She was found by her husband and mother a few days later; she was able to tell them what happened before she died three days later.
- Eleven-year old Aye Pong, her older brother Nya Mon, and a female friend of theirs returned to their village to get their bullock and cart, after they had been forcibly relocated. They caught the bullock and were harnessing it to the cart when SLORC troops arrived, shot Nya Mon dead, then raped the girl and woman. They killed Aye Pong and forced the other woman to act as porter to them.

When soldiers enter a village, they commit violent acts at will, with no reason or provocation.

- Nan Pang, 28, was questioned by SLORC soldiers who asked her to identify the village headman's house. When they realized she was alone in her house, two of the soldiers raped her. After she threatened to tell their captain, a lieutenant beat and kicked her in the chest. Her brother-in-law took her to a hospital in Thailand, but she died anyway.[9]

As elsewhere, many times women were assaulted in areas which were battle zones, when men in the village were away fighting or hiding from government troops, or after male relatives had been forced into working for the army.

- A Muslim mother of five from Rakine state was raped in the evening when soldiers who had seized her husband earlier in the day returned.
- Twenty-five year old Mi Htaw went to the military post near her village to try to see her husband who had been arrested for allegedly having contact with an insurgency group. While her husband watched, she was repeatedly raped.
- In a village in Karen State, where only women and children remained, SLORC troops entered the village and spent the night looting every house. They gang-raped several women including a 34-year old Mon named Mi Yin Kyi; seven others were reported raped at gunpoint in front of the others.[10]

Women also risk sexual assault when they are forced to become porters and labourers at military camps. They may be forced to sleep with the soldiers on a nightly basis, or attacked at random.

- "At Yebyu camp, I was made to dig bunkers, latrines, look after the vegetable garden, fetch water for them, and clean their uniforms," reported a 55-year old Muslim woman.... At night we had to sleep in the same place with the soldiers. The young women were forced to sleep with the soldiers every night."[11]

Soldiers also assault women who simply cross military check points, finding excuses to first detain them. A common routine is for a soldier at one checkpoint to plant suspicious items such as bullets, for example, in the bicycle carriers of young women. At the next checkpoint, a soldier finds the bullets, detains and then rapes her.

- "In Kyanun Zut Village...troops used a similar trick, hiding some bullets at the base of a pot of drinking water for travelers in front of their checkpoint. Then when a pretty girl came past and stopped, the soldiers went out, 'found' the bullets and detained and raped her for three days."[12]

Even when women have escaped to apparent safety in refugee camps in Bangladesh or Thailand, they risks threats of abuse by security forces.

* In April 1993 several Bangladeshi policemen approached the shed of K.K.B., a resident of the Balukhali II camp. Three of them forced her into the woods near the camp. Each of them raped her. Afterwards, they pulled out their knives and threatened K.K.B. that they would use their knives on her if she told anyone what had happened.[13]

Like many of their counterparts in traditional societies, the women of Myanmar also face deep shame and rage after being sexually abused. "...cultural inhibitions linked to subjects such as sex and the serious implications of rape and assault on women compound the problem of lack of outlets for expression and foment deep hatreds."[14]

In their 1999 reports, the Asian Human Rights Commission and Human Rights Watch report continued atrocities as the decade ended. The government continues to control strictly the flow of information in and out of the country, refusing access to foreign human rights monitors and help for the victims of violence.

Kashmir: Claimed by two nations

Violence in India's only Muslim-majority state stems from 1947, when both India and Pakistan gained their independence and each claimed the state known formally as Jammu and Kashmir. The issue was complicated by the fact that Kashmir had been an independent princely state, not part of British India, and preferred to retain its independence. Fighting between India and Pakistan only ended with UN intervention and the 1948 ceasefire line continues to be monitored by the UN.

However, all parties had agreed that Kashmiris should be largely independent. In fact, the 1948 security council resolution called for a plebiscite allowing Kashmir to choose between joining India or Pakistan. The vote never happened and the state remains divided into zones of Indian and Pakistani control, with a third slice bordering Tibet now occupied by China. Pakistan sees Kashmir as the unfinished business of partition, India as an integral part of its territory. Twice, in 1947 and again in 1965 the countries have raged war due to Kashmir.

Today India has some 400,000 troops stationed in Kashmir; nearly a dozen insurgent groups are fighting against Indian rule. India views the insurgency as a surrogate war with Pakistan, which provides militants with money and weapons. Pakistan says it wants the long-promised plebiscite.

But, a vote does not provide a simple solution either, given the fact that the Kashmir population mix is split between Muslims in the Kashmir Valley, a Hindu majority in the Jammu region, and Buddhists in the Ladakh region. Hindus and Buddhists would probably vote for union with India, and the Muslim majority for union with Pakistan. So, it is likely that dissent would continue. In fact, a 1995 poll indicated that many Kashmiris want to join neither Pakistan nor India but still prefer independence.

Militant activities increased substantially beginning in the late 1980's, exacerbated by India's poor governing of the state. Corruption and vote-rigging were factors that led to the outbreak of violence in 1989. Since then both sides have showed similar brutality toward civilians.

Human rights violations have risen dramatically, facilitated by laws that provide security forces with virtual immunity from prosecution. Hundreds of civilians have been executed, not during battle, but because they are suspected of aiding one side or the other. Since 1989, Amnesty International estimates that 17,000 men, women and children have died from violence perpetrated by both Indian government troops and armed opposition groups.[15]

In 1990 the central government imposed direct rule on the state. From the outset, the Indian government's campaign against the militants was marked by widespread human rights violations, including the shooting of unarmed demonstrators, civilian massacres, and summary execution of detainees. Militant groups stepped up their attacks, murdering and threatening Hindu residents, carrying out kidnappings and assassinations of government officials, civil servants, and suspected informers, and engaging in sabotage and bombings. With the encouragement and assistance of the government, some 100,000 Hindu Kashmiris fled the valley.

In mid-1990s Indian forces began arming and training local auxiliary forces made up of surrendered or captured militants. These state-sponsored paramilitary groups have committed some of the most serious human rights abuses; with human rights defenders and journalists among their principal victims.

Brutality on all sides

During the last decade, "Rape has been systematically used as a means of punishing women suspected of being sympathetic or related to alleged militants and as a weapon in the security forces' efforts to intimidate and humiliate the local population."[16] Rape and sexual abuse have been widely reported, despite the stigma associated with sexual assault and its occurrence in remote places.

One of the most notorious cases of gang rape occurred in Kunan Poshpora village in February 1991, when at least 23 women, aged 13 to 80, were reported raped at gunpoint by soldiers of the 4th Rajput Rifles. The government tried to discredit reports of the rape, including medical reports conducted on 32 of the women, and attempted to absolve the army of guilt, accusing militants of orchestrating a massive hoax to make Kashmir an international human rights issue.

Three years later an Indian NGO, Women's Initiative, reported on its visit to the village. Two of the raped women, one with six children, had committed suicide. One, nine months pregnant when raped, delivered a baby whose left arm was fractured; another pregnant women delivered a stillborn. Many were still receiving medical treatment for injuries sustained during the rape. "No marriage had taken place in the village in the last three years. All girls, raped and non-raped, are single. All the married raped women have been deserted. After intervention by militants and elders, two husbands did take their wives back, one on the condition that there be no conjugal relations, the other that he live in the city away from his wife."[17]

By 1995 Indian forces had driven many of the insurgent groups out of Jammu Valley, and began to focus on the rough terrain of foothills and border districts. They engaged in brutal operations reminiscent of those in the Kashmir valley in the early 1990's. Whole villages were surrounded. Residents were beaten and abused, women raped, young men suspected of being militants detained and tortured.

A common practice is to assault villagers whom they believed to have supported militants, or as a means of terrorizing them so that they would not.

- S, a woman of about fifty resides in the foothill area of Doda. In 1998, the Eight Rashitriya Rifles came to her house and took her, her husband and 8-month old grandson to their base. "They began beating me... they used electric shocks on my feet. I was raped. They stripped off my clothes and said they would kill me. The captain raped me, keeping everyone else outside. He told me: "You are Muslims, and you will all be treated like this."[18]
- Residents of another village in Doda told of the army cordoning off about 20 villages in the area for 15 days, and taking some of the local women to the army camp. "They are looking for the militants. But they are unable to find any. So they harass the local population...Our womenfolk are taken into the army camp, all separately. They round up the women, then take two or three in the evening. They come back after two or three days.

They are very shy then, and don't want to talk about what has happened to them. The army has pressured them not to speak about what happened."[19]

Rape to humiliate the community

Rape often occurs during cordon-and-search operations when men are held for identification in parks or schoolyards while security forces search their homes. Often soldiers punish the civilian population at the same time: rape is used to target women who are accused of being sympathizers as well as to punish and humiliate the entire community.

During search operations in a village outside of Shopian, at least six and probably nine women, including an 11-year old girl and a 60-year old woman were gang-raped by soldiers. The doctor who examined seven of the women told Physicians for Human Rights/Asia Watch. "All of the women were weeping. They told me that 'something bad' had happened at about midnight, that 25 army men had come into the village and into their homes. They told me that the soldiers had accused them of feeding and sheltering the militants..."[20] The doctor's examinations showed the women had bruises and abrasions, and positive semen tests.

- S, a woman of about 24, was at home, in the house owned by her father-in-law, about 70, and his wife. There was knocking at the door and three soldiers entered asking for the womenfolk. According to S, "One soldier kept guard at the door and two of them raped me. They said, 'we have orders from our officers to rape you.' I said, 'You can shoot me but don't rape me.' They were there for about half an hour. Two raped me and two raped H (her sister-in-law). Then they left."

- A and N lived nearby and were asleep when 8 or 9 soldiers came to the house about midnight. Four of them entered the house and ordered their father and brother to be taken away; they then entered a room where the women were sleeping. "They did not say anything when they came in...they covered my eyes and mouth with cloths and told us to lie down." A and N were then raped by each of the soldiers.

* G tells of three soldiers entering her house and taking her husband outside. One came back into her home. "He told me, 'I have to search you.' I told him women are not searched, but he said, 'I have orders,' and he tore off my clothes and raped me."
* After militant attacks on security patrol in the village of Bakhihar, soldiers entered the village of Gurihakhar. B, 35, and R, 25, both reported rapes to Asia Watch. A mother of a 13-year old girl provided an account of the rape as if she, and not her daughter had been raped, to protect the girl from public humiliation. A fourth women, who had just given birth in August, was said to have been raped but was too distraught to speak to anyone.

Those who document incidents of rape have also been abused.

* Dr K, a surgeon at a district hospital, was arrested after he arranged for a gynecologist to examine 7 women who alleged they had been raped by security forces who had broken up a wedding and raped all of them, including the bride. He was blindfolded and taken to a military camp, along with two friends with him at the time of arrest. He and his friends were beaten with canes and a metal belt, and detained for four days.[21]

Today sexual assaults on Kashmiri women continue. Sadly, they are part of a pattern of serious human rights abuses by both the Indian government and militant groups that have been "a critical factor in fueling the conflict that is often overlooked".[22]

These abuses have fed the climate of distrust and hatred that exists today in Kashmir.

Notes

[1]"The Intolerable Status Quo: Violence Against Women and Girls", Charlotte Bunch, in UNICEF 1997 Annual Report.

[2]"Integration of the Human Rights of Women and the Gender Perspective: Violence Against Women", Addendum: Communication to and from Governments, January 1999.

[3]Compounding the plight of women in Nepal is the huge problem of their abduction and forced prostitution in Indian brothels. Human Rights Watch estimates that more than a million Nepali women remain, against their will, in these brothels.

[4]"Cambodian Women in Conflict Situation", Keo Keang, in *Common Grounds: Violence Against Women in War and Armed Conflict Situations*, Asian Center for Women's Human Rights, Quezon City, Philippines, 1998, pp. 282–286.

[5]"Women in Afghanistan: A Human Rights Catastrophe", Amnesty International, 1995, p. 1.

[6]As above. Case studies which follow are found on pp. 5–10.

[7]"Integration of the Human Rights of Women and the Gender Perspective: Violence Against Women", written statement submitted by Human Rights Watch, United Nations Economic and Social Council, Commission on Human Rights, January 1999.

[8]"Myanmar: Atrocities in the Shan State", Amnesty International, April 1998, pp. 6, 8.

[9]As above, p. 9.

[10]"Violence Against Burmese Women by the Military", Iyori Naoko, in *Common Grounds*, pp. 288–289.

[11]As above, p. 289.

[12]As above, p. 290.

[13]The Human Rights Watch Global Report on Women's Human Rights, August 1995, Chapter 2, "Sexual Assault of Refugees and Displaced Women", p. 117.

[14]"Situation of human rights in Myanmar", Report of the Special Rapporteur on Myanmar, January 1999, p. 13.

[15]"Summary of Human Rights Concerns in Jammu and Kashmir", Amnesty International, 1995, p. 1.

[16]"India: Torture and Deaths in custody in Jammu and Kashmir", Amnesty International, 1995, Chapter II: Torture, Section 2: Rape and Sexual Abuse.

[17]As above.

[18]"India: Behind the Kashmir Conflict", Human Rights Watch, July 1999, p. 11.

[19]As above, p. 12.

[20]"Rape in Kashmir: A Crime of War", Asia Watch/Physicians for Human Rights, rest of cit to come.

[21]All of the above case material comes from "Rape in Kashmir", cited above.

[22]"India: Behind the Kashmir Conflict", p. 1.

INDONESIA: A HISTORY OF
CONTROL AND CORRUPTION

"After years of authoritarian rule, victims, witnesses and human rights defenders in Indonesia have very little confidence in the criminal justice system of the country. As a result, cases are not reported and the police and prosecutors conclude that there are no cases. Nothing could be further from the truth."[1]

Radhika Coomaraswamy,
Special Rapporteur on violence against women

Indonesia is a complex, ethnically diverse nation stretching nearly 5000 kilometres from east to west, and encompassing nearly 1000 populated islands including Java, Sumatra, Sulawesi, Bali and portions of Borneo and Papua New Guinea.

For almost 350 years Indonesia, then known as the Dutch East Indies Islands, was tightly controlled by The Netherlands, one of the major colonial trading powers of the 17th and 18th centuries. The Dutch drew tremendous fortunes from the agricultural resources of their far flung colony.

During their three-century rule, the Dutch did little to promote the long-term well-being of Indonesian society. They maintained a rigid social structure, building on the existing Javanese caste system; demanded increasing amounts of agricultural yields from farmers; and largely ignored popular education. By 1940 ninety percent of the population was still illiterate.

However, by the early 1900's the expansion of Dutch business, finance and government in the colony created a demand for more local people to be educated to do administrative work. Higher education became available for the first time and a small educated elite developed. These young people were exposed to western political and social traditions and, as a result, became very critical of the injustices of the colonial system.

Out of this group developed the Indonesian nationalist movement. By 1927 an Indonesian National Party, under the charismatic leadership of ex-engineer Achmed Sukarno, was agitating for independence. After World War II, in 1949, the Dutch finally capitulated to both internal and external pressure, and turned over the former East Indies to a new Indonesian Republic under its first president, Sukarno.

But the new Republic was unstable, with no national income, few civil institutions, and violence propagated by religious extremists, regional secessionists, and communist groups. To halt the chaos, Sukarno in 1956 proclaimed a policy of "guided democracy" which meant he increasingly concentrated power in his own hands and, over time, began to fraternize with the Communists.

In 1965, a surprise military coup led to widespread bloodshed: six top generals were abducted and murdered, nearly half a million people died in mass political murders, the Communist Party was obliterated and the government bureaucracy purged of "suspicious" elements. There is still debate as to who planned and led the mysterious coup, but when it was over Sukarno was left with little power. The Army under a previously unknown General Suharto took control and established a highly authoritarian government that ruled for more than three decades.

Suharto's rule promoted economic growth, improved living conditions for average citizens, and supported education and population control. However, it also centralized control, encouraged a powerful military, and allowed widespread corruption among public officials, most notoriously Suharto and his family who are alleged to have siphoned off billions of public dollars.

His rule was also characterized by tight control of regional, ethnic and religious differences within one of the most diverse nations in the world. Indonesia under Suharto was a simmering stew of frustration fed by the central government's exploitation of the resource wealth of Indonesia's provinces.

The 1997–1998 Asian economic crisis was the catalyst for political and social upheaval. It spurred student demands for political reform, and eventually led to the resignation of Suharto after 32 years in power.

Once the strongman was gone, and new president B.J. Habibie appeared to encourage real democratic dialogue, old divisions quickly surfaced. In the euphoria after Suharto's departure, virtually anything seemed possible. When the Habibie government took few actions to back up its rhetoric, disillusionment was intense. Revolt against the status quo grew stronger, partly due to more political openness, partly based on the belief that Indonesia was now in the public eye and could count on stronger international support.

When social discontent bubbled to the surface in the most aggrieved provinces — Aceh, East Timor, and Irian Jaya — the military struck back, as they had in years past, with brutal force and virtually unlimited powers. Despite continued allegations of serious violations of human rights, the Habibie government appeared to support the military — a stance which spurred nationalist movements even more.

Then riots in the capital itself, Jakarta, brought human rights abuses to the doorstep of Indonesia's elite, in the form of assaults on their ethnic Chinese neighbours.

Today Indonesia remains fragile. A new, democratically elected government took control in October 1999, led by President Abdurrahman Wahid (better known as Gus Dur), and is attempting to manage the pent-up frustrations of 30 years. There are those who fear Indonesia will break up into warring fragments, like Yugoslavia. Others are intent on keeping the island nation together, whatever the price. The rest of the world waits and watches as its fourth most populous nation tries to heal its wounds.

Many of those wounds are borne by Indonesian women who have often been the target of the all-powerful military. A few are just beginning to come forward and tell their stories, says the UN Special Rapporteur on violence against women, Radhika Coomaraswamy.

"After years of authoritarian rule, victims, witnesses and human rights defenders in Indonesia have very little confidence in the criminal justice system of the country. As a result, cases are not reported and the police and prosecutors conclude that there are no cases. Nothing could be further from the truth. There are many cases of rape and sexual violence, but people do not come forward. They are terrorized by intimidation and threats from anonymous individuals. They also lack confidence in the system and therefore feel that reporting crimes of this nature is a dangerous waste of time."

After a visit to Indonesia in late 1998, Ms. Coomaraswamy concludes: "...rape was used as an instrument of torture and intimidation by certain elements of the Indonesian army in Aceh, Irian Jaya and East Timor...torture of women detained by the Indonesian security forces was widespread... Among the methods of torture that were employed were rape of the detainee...."[2]

The stories that follow document some of the cases made public thus far. They tell of a military that operated at will against civilian populations, intimidating communities by hitting at their most vulnerable members, women and children. They tell of rape, sexual mutilation, and sexual slavery. And, they are the tip of an iceberg, a fraction of the stories of violence that remain to be told, snapshots in the volatile drama that is Indonesia today.

Aceh: Fiercely independent

Aceh's location on the northern tip of Sumatra, at the head of the Straits of Malacca, has made it an important center of trade and commerce for the last thousand years. Aceh reached its peak in the 16th and 17th centuries when its capital, Banda Aceh, was a major international commercial center attracting traders and immigrants from India, China, and the Middle East. As a result, the Achenese population is a blend of Indonesian, Arab, Tamil, Chinese, and indigenous groups.

For most of its history, this resource-rich province was an independent kingdom ruled by a long line of sultanates. Only in 1903 did the Dutch take control of Aceh, and only after decades of fierce resistance. Aceh was a leader in Indonesia's nationalist struggle for independence, and then fought unsuccessfully to make the new republic a Muslim state. Instead, the central government granted the province the status of a Special Autonomous Territory where Islamic law applies.

Under the Suharto regime, Aceh grew increasingly resentful as it saw the central Javanese-dominated government take a large portion of the revenues from the development of the province's natural resources. In 1971 Mobil discovered massive natural gas reserves in North Aceh and developed a plant that became the biggest in the world, supplying 30% of Indonesia's oil and gas exports.

By the late 70's resentment became rebellion in the form of an independence movement called Aceh Merdeka, or Freedom for Aceh. Though the Indonesian military quickly suppressed the movement, some of its leaders escaped and stoked the fires of independence from abroad. It surfaced again in the late 80's, once more over the issue of the central government's control of the province's resources. Exiled independence leaders, some of them trained by Libya, returned and began attacks on soldiers and non-Acehnese Indonesians who had migrated to the province.

The Indonesian army responded with tremendous force, "killing more than a thousand civilians, often leaving their mutilated bodies by the side of roads or rivers. Many more were arrested, tortured, and arbitrarily detained for months, sometimes years. Hundreds of men disappeared. Many women whose husbands or sons were suspected of involvement with the guerrillas were raped".[3] Between

May 1990 and August 1998, Aceh was designated a military zone or "DOM", which allowed the military to operate with impunity against both insurgents and civilians. In the three districts most affected — North Aceh, Pidie and East Aceh — virtually every family was scarred by the violence of those years.

An inquiry conducted by the local government in early 1999 concluded that in North Aceh alone some 430 people had died in 1989–92, with another 320 still missing. Little data could be found on the reported hundreds of women who were raped and sexually abused during that time.

During a visit to Indonesia in 1999, the UN Special Rapporteur on Violence received reports of large-scale sexual violence during counter-insurgency operations in 1990–91.

- At 2 am one morning more than 20 soldiers came to F's home looking for her husband. They broke down the door and interrogated her and her children. The soldiers left after she told them her husband had gone to his parent's house because they were ill. About an hour later, three of the soldiers returned, questioned her again, and, when she tried to escape, began beating her. Finally they gang raped her, despite the fact that she was six months pregnant.[4]
- N's husband was taken away by soldiers from Kopassus, elite army commandos, and was missing for several days. During that time, he suffered torture that left him deaf in one ear and limping from a fractured thigh. Afraid of being arrested again, he left and went to another village to work as a farm labourer. Kopassus was convinced he had joined the guerrillas, so they took N to the military post and questioned her for 15 days on her husband's activities. On the 16th day they began to use torture: she was undressed and raped by one soldier while the others watched and laughed. Then she was given electric shocks in her ears, nose, breasts and genitalia. Five days later she was released. She has severe internal injuries as a result of the torture.[5]

When former president Suharto resigned in 1998 and the political climate opened up, information on atrocities committed in Aceh began to pour out of the province. Expectations began to build that the government would act to remove the military, prosecute soldiers, and compensate victims. Many observers believed the new Habibie government response fell far short of expectations.

In August General Wiranto, then commander of the armed forces, apologized for human rights abuses and announced the end of military operations. However, an August 31 ceremony marking the military's withdrawal turned into a riot as the angry crowd went on a rampage. "While the initial stoning (of soldiers) may have been spontaneous, there is some evidence that the violence which followed involved local officers, unhappy at leaving extra budgetary sources of income, such as illegal logging and marijuana cultivation."[6] Some 45 civilians were reported killed and more than 100 wounded.

When nothing more had been done by January 1999, the Acehnese were ready to press their demands further. A student congress proposed a referendum on the province's political status; the idea caught the mood of the people and quickly gained support from activists and local officials.

Simultaneously, insurgents who had fled Aceh began to return and the independence movement gained strength. Human Rights Watch acknowledges that the Free Aceh movement has committed its share of human rights abuses, but in an August 1999 report goes on to say: "those abuses pale beside Indonesian army and police excesses. It's as though all the latter know how to do is open fire, and as the casualties mount, so does support for GAM (the Free Aceh movement)."[7]

In September 1999 the Indonesian Parliament passed a bill giving Aceh more autonomy but it appeared to be too little too late. In November 1999 an independent commission established to investigate abuses in Aceh reported to Parliament that it had found 5,000 cases of human rights abuses, including summary execution, torture, and rape and sexual violence. Legislators also questioned several top generals over their role in the 9-year anti-guerrilla military operation in which an estimated 2,000 people were killed.

By December, Parliament was pressing the government to implement a human rights tribunal to try those suspected of committing human rights abuses in Aceh, saying that the government faced a crisis over growing separatism in both Aceh and Irian Jaya. In January President Wahid paid a brief visit to Aceh. He urged a peaceful settlement to the separatist violence but ruled out an independence vote in Aceh, similar to that in East Timor.

In October of this year, as this book goes to press, the situation in Aceh remains unresolved. Many observers believe it to be one of the most serious problems facing Mr. Wahid's government.

East Timor: Decades of military control

Timor, a large island in the eastern part of the Indonesian archipelago, had been divided for centuries between the Dutch, who colonized the west, and the Portuguese, who held the eastern part of the island for more than 450 years as a lucrative source of sandalwood and cinnamon. West Timor became part of the new Indonesian Republic in 1949.

East Timor abruptly gained its independence from Portugal in 1974 after a military coup in Lisbon led to a socialist government which quickly set out to divest themselves of Portugal's possessions. The East Timorese had little time to develop political consensus; as a result, rivalry between three main political groups led to violence. Indonesia, fearing the East Timor example would encourage separatist elements in the Republic, used the violence as an excuse to launch a brutal invasion late in 1975.[8] Six months later Indonesia officially annexed East Timor as its 27th province. The United Nations never recognized Indonesia's right to rule East Timor, though a number of major nations did, among them the United States and Australia, largely because it served their economic and political interests.

For more than two decades, the East Timorese lived with a heavy military presence, while the main independence group, Freitlan, continued to fight a guerrilla war from the rugged hills of the province. In a pattern similar to other provinces, the Indonesian government policy was to repress the local people and culture, largely a mix of Papuan and Portuguese; encourage "Indonesianisation" by promoting migration from other, more crowded areas of the archipelago such as Java, Sumatra and Sulawesi; and exploit natural resources such as coffee and timber, with the military enjoying a near-monopoly over the lucrative export trade.

The military became a law unto itself in this remote province. Organizations such as Amnesty International and the United Nations have documented major, continuing abuses of human rights including arbitrary detention, torture, unfair trials, and political killings and disappearances.[9] At least 100,000 people, out of East Timor's pre-invasion population of 700,000, are estimated to have died as a result of military action. Some observers say the figure is closer to 200,000 if one includes all those who died as a result of the starvation and disease which accompanied the military occupation.[10]

In the early 1990's security forces were still a sizable contingent: 15,000 heavily armed Indonesian soldiers, an equal number of police, and a large number of plain-clothes spies.

The last decade began with one of the most serious and well-known massacres of civilians in November 1991 at the Santa Cruz Cemetery in the capital of Dili. The day started with a religious service commemorating the death two weeks earlier of a young political activist, killed by Indonesian security forces in a church where he and a number of other activists had taken refuge. It was followed by a peaceful pro-independence march to the grave, with the display of separatist banners and slogans. Two soldiers — a major and a private in civilian dress — were reported to have been injured along the route.

News of the injured soldiers were relayed to troops and when the crowd reached the cemetery, soldiers opened fire. The cemetery walls and large crowds made it difficult to escape. According to eye-witnesses, those wounded and supposedly being taken to a military hospital were deliberately killed during transit or at the hospital's morgue. An estimated 270 people were killed that day.[11]

Far less well-known are the assaults against East Timorese women that were a consistent theme of the Indonesian military occupation.[12] Often girls and women were assaulted because they could not or would not comply with a request from the military, whether legitimate or not.

- In October 1993, soldiers went from house to house in Luwa village to force the women to accompany them to a party in another village. One of the women, Ms. B, refused the request and was then beaten until she bled, and stripped naked in front of her family, Another, Joana Soares, was gang-raped, then stabbed to death.
- In January 1994 Mrs. "A" from Bua Narak hamlet was arrested and raped by a platoon commander when she attempted to travel to a nearby village. Her inability to show a travel document was the pretext used to accuse her of being a guerrilla, arrest and assault her.

At other times, there was no pretext at all for attacks, just the certainty of occupied troops that they had the power to do as they pleased.

- A private in the 612th battalion simply walked into the house of Ms. "A" carrying an M16 rifle and threatened to shoot her parents if she resisted. He returned several times to rape her and she became pregnant as a result.
- Mr. X told of soldiers abusing as many as 50 women in his village. His younger sister, then 25, was raped by an Indonesian soldier and subsequently became pregnant and gave birth to a baby who was 10 months old at the time of the interview.[13]

Another common practice was to force women to become so-called "local wives" or sex slaves for the occupying forces.

- Dukai (not her real name), now 39, was abused and raped by successive Indonesian soldiers during the period from 1977 to 1991. As a result of the rapes, she has five children, ranging from primary school age to the early 20's. She suffered not only from the abuse, but from outright rejection from her own people since she was considered to have "served" the military. Today she lives with her uncle and aunt, doing odd jobs to earn enough to pay for her children's education.[14]
- Odilia Victor, an East Timorese woman who escaped to Australia, told of her sister who had been held as a sex slave in Dili for about a year, after her husband fled to the bush. She was then forced to become the local wife of the Indonesian Air Force officer whose house was next door to her family home. Even so, she never dared to try to contact her family, nor her family her. Odilia explained that many women singled out to be local wives or sex slaves have little education, and therefore are especially vulnerable to intimidation.

Odilia also told of the wives of guerilla leaders, left behind in towns and villages, forced to live with Indonesians or East Timorese siding with the military. This allowed monitoring of any communication with their husbands. It also contaminated the women in the eyes of their own community...thus furthering the military objective of weakening the unity and morale of the enemy.[15]

Violence in East Timor increased during 1999 as then-President Habibie announced early in the year that he would give East Timor the choice of autonomy under Indonesian sovereignty or independence. A network of 13 district-level militias sprang up, most likely supported by the Indonesian army, aimed at intimidating Timorese into staying as part of Indonesia.

Violence worsened even more after the September 4 announcement of the referendum results, when nearly 80% of the population voted for independence. For two weeks militia groups rampaged across the province, killing and destroying property. More than 150,000 East Timorese were moved into camps in West Timor, reportedly forced to leave by the Indonesian military and the militias. Women in Dili were raped and sexually harassed by both militia and Indonesia military. Militia also reportedly raped many women on a boat that was taking displaced persons from Dili to West Timor. Other reports indicated that women were being repeatedly raped in camps in West Timor. After two weeks of destruction and terror, as well as intense diplomatic pressure, President Habibie agreed to allow UN peacekeepers to enter East Timor and help restore order.

Since then, new president Wahid has supported the East Timor human rights inquiry launched by Habibie; he has gone further to pledge that officers will stand trial if there is enough evidence to indict them. In February 2000 Wahid suspended former army chief General Wiranto from government, after he and five other officers were named by the Indonesian inquiry as being ultimately responsible for the terror that swept East Timor after the referendum. Indonesian leaders are urging that the country's own inquiries be completed quickly to head off the possibility of an international war crimes tribunal on East Timor which had been considered by the United Nations.

Today East Timor is in the midst of a still-precarious transition. The United Nations is charged with helping to prepare the territory for independence. It has estimated that at least $3 million is needed to rebuild the small nation's infrastructure, destroyed during months of fighting: the capital of Dili, for example, is desolated with rows of blackened shells where homes and businesses once stood. East Timorese leaders say that more than 100,000 of their people are still being detained in West Timor, where army-supported militias retain control.

Even the most optimistic projections put full independence at least two years out, given East Timor's current lack of money, infrastructure, skills and equipment. And, even full independence will still mean that East Timorese will live uneasily with Indonesian West Timor, plagued by the distrust and hatred built up during decades of unpunished brutality.

Irian Jaya: Often overlooked troublespot

Irian Jaya occupies the eastern half of the island of New Guinea, the world's second largest island which lies in the Pacific Ocean between Asia and Australia. It is the most remote and undeveloped of Indonesia's provinces, making up one-fifth of Indonesia's land area but serving as home to less than one percent of its population. Its landscape is wild and mostly undeveloped, marked by dense jungle and a high central mountain range. The indigenous people, comprised of Negroid and Melanesian races who settled the island in Neolithic times, share little with most other Indonesians in terms of history, culture, or religion.

When Indonesia gained independence in 1949, the Dutch succeeded in retaining the land then called West Papua, which they saw as the last bit of their former empire — and a valuable political asset if the new Indonesian Republic did not hold together. They openly encouraged nationalism, prepared the colony for self-rule, and set 1970 as the date for independence.

However, under heavy pressure from the US, which supported the new Indonesian government's claims, the Dutch turned West Papua over to an interim UN administration. Indonesia sent troops into the territory in 1963 and laid the groundwork for an "act of free choice" in 1969.

Before the 1969 vote, then president Suharto waived the agreed-on referendum, hand-picked tribal representatives, and brought them together to vote under close Indonesian military supervision. The result was predictable: all agreed to join the Indonesian Republic.[16]

In the nearly four decades since then, both violent and non-violent movements within Irian Jaya have continued to work for independence. Nationalism has been fueled by Indonesia's treatment of the Irianese, which resembles that of an occupying colonial power.

- The government has exploited the area's considerable natural resources — the world's richest copper deposits, some of

Indonesia's largest oil fields, and important gold, uranium and timber resources. Yet, as in other provinces, most of the wealth has been funneled back to the central government in Java.

- Indonesian authorities have refused to recognize indigenous land use and ownership, ruling that land can be considered "unused" and thus subject to government claim if it is not being actively cultivated. This is in spite of the fact that Irianese may see the land as important as part of a cycle of shifting agriculture, or for gathering of food or medicine. So far over a million hectares of swamp and rain forest have been cleared for either commercial use or migrant resettlement, with plans to clear another million in the next 10 years.

- An official "transmigration" policy has resettled in Irian Jaya several hundred thousand Indonesians from more crowded islands, particularly Java and Sulawesi. Both the government and these new residents often view the black-skinned Irianese as inferior. In urban areas, this may result in different rates of pay for the same work.

Since 1969, the military have maintained a major presence in Irian Jaya, claiming it was necessary to protect Indonesian economic interests, including the world's largest open pit goldmine. In fact, the military has responded brutally to conflicts, protest activities, and political dissent, with reports of murder, imprisonment, torture and sexual abuse of Irianese.[17]

The activities of a guerrilla separatist movement, the Free Papua Movement or OPD, intent on uniting Irian Jaya with Papua New Guinea has led to large-scale military action against Irianese villages in an attempt to suppress the guerillas. But, because of the military's tight control, as well as the province's remote location, it has been difficult to gather information to substantiate these reports.

However, a 1998 UN report notes that "sexual violence by the Indonesian security forces in Irian Jaya appeared to be taken for granted, both by the authorities and the local population".[18] The report cites the following incidents:

- In 1987 A, from Jila village, was raped by a soldier while she was working in the fields. When she told her parents, they went to the military post to demand justice but were beaten and sent home. Then her two brothers, one a priest and the other a village chief, went to the post to complain; they were also beaten. The following year A, who had been a virgin, had a child as a result of the rape. "It is alleged that soldiers raped many women in that area. Women were afraid that, if they resisted, their families would be attacked. There are many children as a result of the rapes."
- In February 1996 troops from all over Indonesia are reported to have come to the Mapnduma area, where soldiers are alleged to have raped women indiscriminately. "(G)irls as young as 12 were victims, as were mute, mentally retarded and pregnant women."
- Pro-independence demonstrations in July 1998 were disrupted by heavily armed troops. Reports claim that "women were taken out to sea on Indonesian navy ships, where they were raped, sexually mutilated and thrown overboard. Women's corpses reported washed up on the Biak coast. Some of them showed signs of sexual mutilation; breasts had been removed."

Today the continuing struggle in Irian Jaya is overshadowed by other problems faced by Indonesian leaders. The province — and its aggrieved population — receives little attention from the media, the Indonesian government or the international community.

Jakarta: Targeting ethnic Chinese

The events which led to the resignation of President Suharto on May 21, 1998, also spawned shocking violence in the heart of Jakarta, the kind of violence more typically seen in military controlled provinces such as Aceh or East Timor.

The precipitating factor appeared to be the fatal shooting on May 10, 1998, of four university students who were demonstrating against

President Suharto. Within the next three days, nearly 2000 people were killed and some 168 women sexually assaulted while over 4000 shops and 1,000 homes burned to the ground. Though both Chinese and non-Chinese died in the fires, the target of the riots were Chinese neighbourhoods. Chinese women were also clearly targeted for sexual assault, with rapes primarily occurring in west and north Jakarta where they lived.

But, as Human Rights Watch notes in a 1998 report, this data "cannot adequately convey the terror that many ethnic Chinese experienced. Human Rights Watch met with one young woman who jumped from the fourth floor of a building to escape a mob; both her feet were smashed, and one heel is permanently damaged. Another man had his hands torn up when he climbed over a barbed wire fence holding his three-month-old daughter to escape rampaging youths yelling anti-Chinese chants."[19]

The six-million ethnic Chinese Indonesians, who constitute just under 3 percent of the population, have lived uneasily in their country for many years. Beginning in 1967, the government of Indonesia declared a policy of assimilation with regard to their Chinese citizens: Chinese language schools were closed, Chinese festivals and religious festivals were no longer celebrated publicly, ethnic Chinese were asked to take Indonesian names and to carry identity cards showing their Chinese origin.

The Suharto era of shadowy influences, outright corruption, and injustice toward ordinary citizens, fed the perception that the Chinese controlled the Indonesian economy in collaboration with Indonesian power elites. While many Chinese are very successful in business, the UN Special Rapporteur reports that many of the victims of the May riots whom she interviewed came from lower middle-class backgrounds, some of them single women living alone. "It appeared that the victims were in fact, poor ordinary women who had very little 'control of the economy'."[20]

It is generally agreed, by both the Indonesian government and independent observers, that the riots were deliberately instigated by groups of men, arriving in jeeps and on motorcycles, who would incite the people to riot, and then help break into and loot buildings. It is less easy to identify the instigators. Some believe that military close to Suharto were concerned about the growing student protests and provoked violence in the hope that it would be an excuse for the imposition of martial law. Some local criminals have confessed they were paid to riot. The UN Special Rapporteur reports seeing a video of the riots in which members of the armed forces stand by and watch the rioting and looting.

Despite initial denials by the government that systematic rape had been part of the planned violence, local volunteer groups, notably Tim Relawan (Volunteer Team for Humanity) painstakingly gathered details from eyewitnesses and rape victims too frightened to report their assault to anyone associated with the government. They identified the location of the rapes, establishing that they were concentrated in West and North Jakarta, and outlying areas. In its reports, Tim Relawan also establishes that the sexual assaults and riots occurred simultaneously.

All of the following data comes from the Tim Relawan report on information gathered between May 13 and the end of July 1998. By that time, volunteers had documented 168 victims of sexual abuse, 138 of them who had been raped. Most assaults took place in the Chinese neighbourhoods of Jakarta, but 16 Chinese women were assaulted in the cities of Solo, Medan, Palembang and Surabaya.[21]

- On May 13, the home of mother and daughter W and L, aged 50 and 26, was attacked and plundered by a group of unknown men. They forced the son of the family to rape his sister, and a servant to rape the mother. Both were then raped by others in the group. The home was then burnt, and both brother and sister thrown into the fire. The mother committed suicide by jumping into the fire herself.
- On May 14, N and L, sisters aged 19 and 21, were raped by 7 to 10 men, some of them yelling: "Because you are Chinese,

you are raped." Their shophouse was burned and the sisters tossed into the flames. The parents were forced to witness the tragedy.

- Again on May 14, a family's shophouse was entered and looted, and the three sisters at home, R, L and M, assaulted. R was stripped of her clothes and forced to watch her sisters being raped, then thrown downstairs where a fire had started. L and M died, but R survived because someone helped her escape.
- Public humiliation and sexual abuse just short of rape were also widely reported. In one incident a group of about 15 men entered a Jakarta bank where 10 ethnic Chinese women were taking refuge from the riot. The men locked the door, ordered the women to take off their clothes and ordered them to dance. In a similar incident in Medan, female students were stopped by police as they tried to flee campus violence. The officers forced them to take off their clothes and perform callisthenics. In both cases, the women reported being fondled but not raped.

As horrifying as these attacks are the threats and assaults against victims who survived the attack, and those who then attempted to help the victims or to report on the violence.

- "Is a grenade not enough? I know where your children go to school. I know what uniform they wear, what time they go to school and what time they come home." (Anonymous telephone caller to a volunteer, June 1998, after a live grenade was found in the volunteer's front yard.)
- "Sorry, I'm not coming...though I do want to talk to you about what happened. Just a few minutes after calling you up, we got a terror call. We are very afraid of having greater troubles..." (Mother of a rape victim to a volunteer, June 1998.)[22]

In October a young women known as Ita, daughter of a Tim Relawan volunteer, was brutally murdered. Police allege the murder was the result of a neighbour attempting to rob Ita's home, but human rights activists think otherwise. Understandably, the murder had a profound impact on the human rights community.

Unfortunately, the initial reports of rape used were met with skepticism by government officials who questioned the credibility of NGOs since no victims had come forward to report to the police. As international shock and pressure from Indonesian women's advocates mounted, however, the government shifted its position and the chairman of the National Commission for Human Rights finally acknowledged that "We can say without any doubt whatsoever that on those dates (May 13–15) mass rape of ethnic Chinese women occurred that was widespread, systematic and sadistic."[23]

As yet, however, there has been no serious investigation of those who planned or carried out the violence.

Notes

[1] Radhika Coomaraswamy, Special Rapporteur on violence against women, "Integration of the Human Rights of Women and the Gender Perspective", Mission to Indonesia and East Timor on the issue of violence against women, United Nations, 1998, pp. 9–10.

[2] As above.

[3] "Why Aceh is Exploding", Human Rights Watch Press Backgrounder, August 1999, p. 2.

[4] "Integration of the Human Rights of Women and the Gender Perspective", p. 18.

[5] As above, p. 3.

[6] "Why Aceh is Exploding", p. 2.

[7] As above, p. 3.

[8] In an interview with Radio Netherlands in 1995 retired Indonesian Lt.-Col Subiyanto told of his experience during the invasion and his subsequent refusal to be sent back to East Timor. He explained that he and some other soldiers refused orders to return to East Timor "Because we read a bit about East Timor. Though colonized by the Portuguese for 350 years, no mass killings ever took place. But after the Indonesian government sent Abri there, Abri massacred the East Timorese...There were so many dead. They were people just like us. Whether you wanted to or not, you were ordered to go in. Then it was too late, we had to defend ourselves...But once I knew a little about the history, I thought If I go I'll just be killing ordinary people. Sure, we

have to follow orders, but you're not allowed to do that." Extracted and translated from the Radio Netherlands documentary 'Tragedi Timor Timur' 1975–1995, part III, December 1995, as quoted by *Inside Indonesia*, Edition 50, April–June 1997.

[9]See, for example, "East Timor: The September and October 1995 Riots", Amnesty International, January 1996; Report of the High Commissioner for Human Rights on the Human Rights situation in East Timor, 1999.

[10]The Report by UN Special Rapporteur on Extrajudicial, Summary or Arbitrary Executions, United Nations, 1994, p. 3.

[11]Briefing on East Timor, Amnesty International, December 1996, p. 1.

[12]Case material from Joint Committee for the Defense of East Timorese, as cited in Common Ground: Violence Against Women in War and Armed Conflict Situations, Asian Centre for Women's Human Rights, 1998, p. 294.

[13]As above, p. 295.

[14]Menyilam Kemarau, Fokuper, Dili, East Timor, 1999, pp. 31–32.

[15]As above, p. 300, based on testimony before UN Human Rights Commission.

[16]See, for example, "Indonesia Alert: Trouble in Irian Jaya", Human Rights Watch, July 1998.

[17]See, for example, "Irian Jaya: Recent Arrests", Amnesty International, May 1996, and "Indonesia: Human Rights and Pro-Independence Actions in Irian Jaya", Human Rights Watch, December 1998.

[18]Report of Special Rapporteur on Violence, Mission to Indonesia and East Timor on the issue of violence against women, p. 19.

[19]"Indonesia: The Damaging Debate on Rapes of Ethnic Chinese Women, Part II, Human Rights Watch", December 1998, p. 1.

[20]Report of Special Rapporteur on Violence, p. 12.

[21]"The Rapes in the Series of Riots", Early Documentation No. 3, Tim Relawan Untuk Kemanusian, Jakarta, July 1998, pp. 5–8.

[22]As above, pp. 9–10.

[23]"Jakarta mass rape was systematic", Kompas Daily News, 8 July 1998, Online version.

86

WHERE WE ARE, WHAT WE CAN DO

"The tragedy of East Timor, coming so soon after that of
Kosovo, has focused attention again on the need for timely
intervention by the international community when death
and suffering are being inflicted on large numbers of
people; and when the state nominally in charge is unable
or unwilling to stop it... From Sierra Leone to Sudan, from
Angola to Afghanistan, there are people who need more
than words of sympathy. They need a real and sustained
commitment to help end their cycles of violence, and give
them a new chance to achieve peace and prosperity."[1]

Kofi Annan, UN Secretary-General

In the preceding chapters we heard the voices of many women
who have suffered terrible sexual assaults in nations throughout
the world, and we have seen the impact of these barbarous acts
on women, their families, their communities.

Such acts of violence lead to many questions, questions asked
by governments and their people as they face an era of great social
and political change. What does it mean to be a sovereign state?
What are the duties of sovereign states toward their citizens? What
are the fundamental rights of every human being, of women in
particular? How do we resolve differences among nations in how
they define the rights of their citizens? When is it legitimate for
other states or the international community to intervene to protect
citizens being abused by their own government?

Seeking answers to these questions must be a central part of any discussion of sexual violence against women during times of conflict. Changes in international politics — the demise of colonial empires and the rise of new independent states, the end of the Cold War and a turning inward among major powers — have greatly reduced the chances of major international conflicts but greatly increased the number of conflicts *within* countries. In Afghanistan and Rwanda, in Burma and Bosnia, in Haiti and Indonesia, we have seen how recent conflicts arise from political, ethnic and religious divisions within societies where legal and other institutions are still in early stages of development. The political and social conditions that continue to spawn violence include:

• Dictatorial states with near-total power over their citizens
• Powerful military forces that make their own rules and act with impunity
• Political leadership that encourages "demonization" of specific ethnic or religious minorities
• Weak or nonexistent legal infrastructure to protect citizens
• Government control of media, universities and other "thought leaders"

Increasingly, the targets are often ordinary people. Nine civilians die for every one soldier killed; more aid workers die than do peacekeeping troops.

Women and girls are a particular target. "There is a deplorable trend toward the organized humiliation of women, including the crime of mass rape," acknowledged former UN Secretary General Boutros-Ghali in an address to the Fourth World Conference on Women held in Beijing in September 1995.

Why sexual assault?

Earlier chapters showed that there are many historical reasons for the unfortunate use of sexual assault as a weapon during times of conflict.

- First, the historical position of women in society has led them to be viewed for many centuries as acceptable targets during conflict — possessions of men and, therefore, both spoils of war and symbols used by enemies to humiliate their menfolk. In East Timor, occupying Indonesian soldiers forced village women to become "local wives" or sex slaves. In Myanmar, women are taken to become porters and labourers at military camps, then raped at random.

- Second, in recent years, there have been more conscious attempts to use women as surrogates for the entire community: Break them, break the community. Recall the answer from the Pakistani politician asked to explain why systematic rape had been part of the effort to quell rebellion in Bangladesh: "Put a gun in (soldiers') hands and tell them to go out and frighten the wits out of a population and what will be the first thing that leaps to their mind?" Recall, too, that the earliest rapes in Bosnia took place before any fighting began, usually in front of the victims' families and neighbours. The goal was to terrorize the population and "encourage" them to flee the disputed territory.

- Third, despite the powerful impact on victims and their communities, rape is almost risk free for soldiers especially since sexual assault has so seldom been punished. "Some military tactics put the soldier's life at grave risk...others entail great physical hardship...many tactics are extremely expensive. Sexual assault is none of these...When the soldier has the option of destroying the armed enemy by fighting or destroying the unarmed female by raping and consequently harming the male by defiling his "property", it is easy to understand why rape has been so prevalent."[2] In Kashmir, even in the most notorious cases of systematic rape by Indian soldiers, the government has done little to punish troops, often seeking instead to discredit the medical reports which prove sexual assault occurred. In Haiti, women hesitated to even report rapes since they feared retribution from authorities sympathetic to their attackers.

We have seen that women are assaulted for the same reasons, in whatever country they call home. Women become victims of violence because:

- They are wives, sisters, mothers of men who are considered "the enemy"
- They are believed to hold "anti-government" political beliefs
- They are easy targets of ethnic and religious hatred
- Their violation will cause fear, panic and retreat within their communities
- They are seen as under the control of occupying military and security forces and property that can be used at the whim of the aggressor
- Their pregnancy and bearing of "enemy children" will cause further disintegration of their communities.

And, we have seen the horrendous impact of this violence on individual women, their families and their communities.

The horrible consequences of rape

Many women are killed after being raped: 80% of women in Bosnian rape-death camps died; an estimated 200,000 or more Rwandan women were murdered after they had been sexually assaulted; women taken from Irian Jaya to Indonesian navy ships were raped, sexually mutilated, and thrown overboard.

If the women are not murdered, they are often very badly injured — injuries that can prevent them from bearing children and cause physical distress for many years. Think of Immanuelle, the Rwandan rape victim who survived the massacre in her country only to live with excruciating abdominal pain because her fear and shame kept her from seeking medical care. When she did finally consult a doctor, she learned that surgery could end her pain, but that she had been infected with AIDS by her assailant. Now she says, "I am alive but this is a living death."

Emotional and psychological pain can be as great a torture as physical pain, or even worse. A middle-aged Bosnian Muslim woman, a lawyer and judge when violence broke out, tells of her continuing terror after two months in a rape camp: "The consequences still appear on a physical and mental level. I can't sleep nights...I still wake up in a sweat thinking of the camp. Those nights when we lay there side by side holding hands, when the Serbs would come and call out a name and take a woman out."[3]

Recall, too, the stories of young Muslim women from conservative traditional communities, who have never even been seen naked by their husbands. Imagine the trauma and despair such a woman feels after being raped by many men for many months. In fact, significant numbers of these women, like 16-year old Nadia in Afghanistan, choose suicide if they have the opportunity, rather than submit to rape.

There is also the desperate silence and isolation such women face when they hide their assault out of fear husbands and families will reject them, even kill them, because of the perceived dishonor associated with rape.

Then there is the social impact of rape which "erodes the fabric of a community in a way that few weapons can. Rape's damage can be devastating because of the strong communal reaction to the violation and pain stamped on entire families...In Bosnia-Herzegovina, Myanmar and Somalia, refugees frequently cite rape or fear of a rape as a key factor in their decisions to seek refuge."[4] Even in a major city like Jakarta, with the social and legal supports it offers, Chinese families fled rather than face the risk of assaults on female family members.

Recall the psychological warfare experts in Bosnia who advised that attacks on women, particularly those of child-bearing age, were the most effective method to "spread confusion among the community thus causing fear and then panic". Think of the Kashmiri village, where a majority of women were raped by Indian soldiers in 1991. Three years later, when aid workers visited the village, they learned that all the raped married women had been deserted, and no marriages had taken place in the village since the attack. Think of the Rwandan women who have been ostracized by their clansmen who fled the massacre, the women accused of giving their bodies to the enemy in order to save their own lives.

These attacks demoralize communities by demonstrating their inability to protect women and tear them apart as victims are stigmatized and rejected. In fact, "If the sexual assault is pervasive, the entire ethnicity can be in danger of being eradicated, particularly in Islamic societies where it is believed that the ethnicity of the father determines the ethnicity of the offspring. Sexual assault can also result in genocide where the social structure is destroyed and the woman becomes unmarriageable or divorceable if married; or where, as a result of repeated rapes, the woman can no longer bear children; or where the result of sexual assault is a destruction of the community."[5]

All of these physical, emotional and social consequences confront women as they struggle to put their lives together in the context of what are essentially wartime conditions: Few medical or mental health resources; ruined infrastructure, including destruction of homes; extreme poverty; disappearance or death of loved ones and, for many, life in refugee camps where they may face continued abuse.

Violence begets action — finally

The international community at last was roused to action by NGO and media reports of the sexual atrocities being committed in the former Yugoslavia during the spring of 1992. Within the space of two years, the United Nations had taken several steps to call attention to the severity of the problem of violence against women.

- **February 1993**. The United Nations charters an international tribunal to investigate human rights crimes in Yugoslavia; the governing statute for that tribunal expressly refers to rape as a crime against humanity.

- **December 1993.** The United Nations Conference on Human rights formally recognizes violence against women, and specifically violence during armed conflict, as a human rights issue. The General Assembly acknowledges an alarming amount of female-targeted violence and adopts a Declaration on Elimination of Violence Against Women. (Until this point most governments tended to regard violence against women as a private matter between individuals and not as a pervasive human rights problem requiring state intervention).

- **March 1994.** The UN Commission on Human Rights appoints a Special Rapporteur on violence against women, with a mandate to collect and analyze data from sources such as governments, specialized agencies, and nongovernmental organizations, and to recommend measures aimed at eliminating violence. One of the three main areas the Rapporteur will address is that of sexual violations during times of conflict.

- **November 1994.** The UN charters an international tribunal to prosecute suspected war criminals who participated in the massacre of Rwandans beginning in April 1994. The statute creating the tribunal lists rape as a crime against humanity, and also refers to rape, enforced prostitution and indecent assault as violations under the Geneva Conventions.

- **November 1995.** The Fourth World Conference on Women takes place in Beijing. In their platform for Action, conference delegates identify women and armed conflict as one of 12 critical areas to be addressed and declare that rape occurring during armed conflict is a war crime.

These are important developments that many observers believe are establishing a path for even more dramatic innovations to protect human rights. The tribunals in particular have helped to strengthen the concept of universal jurisdiction — that is, the right of the international community and unaffected states to prosecute and punish leaders of another state who commit specified major crimes The tribunals also brought women into their process for the first time which "resulted in remarkable strides in the discussion and analysis of women's issues, both inside and outside the tribunal".[6]

Yet even as these steps were taken, the Tribunals were slow to address the real-world violence in the former Yugoslavia and Rwanda. Though the mechanisms were in place, the will to use them was not. In each case, members of the United Nations continued to wrangle over exactly what should be done and how much funding should be allocated.

Again, it took intense lobbying from nongovernmental organizations working in the field to bring sexual assault to the foreground as a crime to be acknowledged and prosecuted. Even so, these statutes maintain the terms and standards of current humanitarian law, with rape and enforced prostitution considered "outrages on human dignity" not acts of violence or torture. They do, however, include rape and forced prostitution in the definition of crimes against humanity, thanks to extensive lobbying from women's groups and other NGOs.

After several years of work, the tribunals have found, prosecuted and sentenced only a handful of those responsible for the violence against women in the two countries.

Sometimes yes, sometimes no

Today we see the role of the international community as inconsistent and contradictory. No longer are former colonial powers willing to intervene in the new nations that once represented their far flung empires. The United Nations is the only organization with broad enough support to take on an international humanitarian role, but as yet it has been unable to get support for a clear and consistent policy with regard to dealing with violence inside states.

As a result, much of the most effective intervention has come from humanitarian nongovernmental organizations: both major international groups such as the Red Cross, Amnesty International, Medicin sans Frontier, and Human Rights Watch, as well as specific regional and local women's and human rights groups who work to document, publicize and intervene in situations where women are subjected to continuing violence.

Some observers see the developed nations of the world looking inward, withdrawing from the international scene. "We have come to fear very much the threat and use of force, even in the most noble causes," says David Malone, a former Canadian diplomat and currently president of the International Peace Academy, an independent group that works with the United Nations to study conflicts and their resolution.

"In the absense of a willingness to take certain risks, we will see a situation in which conflicts are allowed to deteriorate and, in effect, rot until a point where they are extraordinarily difficult to address, and only then will the international community be willing to take forceful action."[7]

An alternative to this bleak scenario, many government and NGO leaders believe, is a permanent International Criminal Court which could move quickly to investigate and prosecute the most egregious aggression: war crimes, crimes against humanity and genocide. The Court would be created by international treaty, not by the UN General Assembly or Security Council; thus, though a body of the United Nations, it would not be subject to bureaucratic and cumbersome UN procedures and rules to the extent that the tribunals have been. As a result, the ICC would have a better chance to operate effectively and efficiently.

The idea for such a court was first mooted after World War II, but Cold War tensions kept major powers from reaching agreement. Again, the atrocities in Bosnia served as a catalyst. It become clear, as it had after Nuremberg, that the global community needed a way to hold individuals accountable for violations of agreed-on international law. Otherwise, the only recourse is to impose sanctions or embargoes, or in rare cases, to take collective military action. Such responses often hurt innocent civilians more than the offending individuals.

In 1994, the UN General Assembly adopted a resolution calling on the International Law Commission, another UN body, to draft an actual statute to establish the court. The so-called Rome Statute has so far been signed by 83 nations, agreeing to the idea of such a court.

The next, much more difficult step is the ratification of the statute, the treaty that will actually establish the Court and determine the details of its mandate, process and procedures. Only four nations have already ratified the statute; 60 are needed to establish the Court. Observers estimate that it will take at least two years, perhaps as many as 10, to ratify the treaty. Again, a nongovernmental group, the Coalition for an International Criminal Court, is the largest single force working to make the Court a reality — it brings together more than 800 NGOs and international law experts.

Would an international court really make that much difference in the end? Many in the field of humanitarian law believe it would.

"Criminal trials even of a few archcriminals, followed by conviction and appropriate punishment serve two principal purposes. They constitute an acknowledgement...of the suffering inflicted on the victims. International prosecution and punishment are particularly significant — an unambiguous statement that the whole world has joined in the condemnation of these criminals...The other purpose served by trials is to demonstrate that the most fundamental rules that make a civilized society possible may not be flouted with impunity and that even the highest leaders cannot be shielded."[8]

Once again, women's groups have been at the heart of efforts to incorporate gender perspective into the new Court. They are working to ensure that all provisions of the Court are nondiscriminatory on the basis of sex and that sex-based crimes are defined appropriately. Specifically:

• Sexual violence in armed conflict should be considered as torture, as defined in international human rights law

- Victims of sexual violence should be able to seek compensation for their injuries
- Sexual violence during armed conflict should be regarded as persecution under international refugee law

Where to go from here?

The proposed Court holds out hope, but the battle to stop the use of rape and sexual assault as weapons of conflict — as well as to gain justice for those who have already suffered assault — is far from over.

What can you do? If you are a reader of this book, then you are educated, aware and capable of having influence on this issue — far more so than the many women whose stories are told in previous chapters. The sum is often more than its individual parts: Your contribution, however small or locally focused, is important; in fact, it is critical.

Keep in mind one of the clearest and simplest lessons we can, and hopefully have, learned from this harrowing tale of sexual assault.

It was not governments or international agencies who took the lead in documenting the assaults, publicizing the violence, organizing people, and mobilizing the political will to reverse centuries of sexual violence that had gone unchallenged and unpunished. It was individuals in countries throughout the world, often as part human rights organizations, who have led the battle — and must continue to do so.

Each of you — female or male, of whatever nation, race or religion — are needed in the ongoing struggle to end these horrific acts of sexual violence. Use this book to determine what you are willing and able to do, whether it be large action or small. Use it as a starting point to educate and motivate yourself and your circle of friends and family; to find ways to lobby your own government; and to support international organizations who are working to stop violence against women. The list of resources following this chapter can help you get started.

Whatever you do, do not turn away from the pleas of the women who have told us their painful stories.

5 things that you can do now

1) **Educate yourself.** Do not simply accept government statements and superficial media stories. Recall how the media were manipulated to incite hatred and violence in both Bosnia and Rwanda. Use the Internet to locate the site of a UN agency, women's organization, or human rights organization which you respect and visit it regularly. Or subscribe to newsletters or periodicals from one such organization and read it regularly.

2) **Take a stand against attacks, subtle or blatant, public or private, on individuals or groups based on sex, race or nationality.** When it occurs within your own community, speak up in appropriate venues: church, school, government meetings, letters to the media. (Remember that in Rwanda, talk show hosts flippantly first talked of the Tutsi women in sexually offensive ways, to plant the idea that they were different, "evil", and therefore not worthy of respectful treatment.)

3) **Take a stand against violence toward women in your own community and region.** Your actions could be as simple as writing opinion pieces for your local newspaper, organizing an exhibition at a local church or school, or lobbying your government, as well as regional organizations, to take the actions recommended later in this chapter.

4) **Join or organize a local or regional group to directly fight for the rights of women closest to home.** It can sometimes be more comfortable to help solve a long-distance problem than one next door. Focus on helping the women in your own country, or neighbouring ones, who are victims of sexual violence.

5) **Support one international human rights organization.** Provide financial support, through regular donations, but also moral and

political support by responding to the organization's request for member action such as letters to political leaders.

5 things that your country should do...and you should lobby for

1) **Ratify the International Criminal Court treaty**. Lobby other nations to do the same.

2) **Take a firm, public stand on acts of sexual abuse which occur during conflicts anywhere in the world**. Every time we fail to stand up to aggression against women, we allow it to spread. Rwandan leaders watched the Bosnians go untouched and gained confidence that no one would stop their violence — and no one did. Individual nations need not wait for the international community to condone human rights violations. Urge your leaders to speak out whenever and wherever sexual assaults take place during times of conflict.

3) **Discipline military and security forces for any act of violence against women**. Military leaders, not only guilty soldiers, should be punished.[9]

4) **Review and amend national and local laws to ensure that they uphold women's basic rights**. Attacks on women during times of conflict are encouraged by the continuing second class status of women in the larger society.

5) **Educate the general population on fair treatment of women and girls**. If a population shrugs or ignores domestic violence, sexual harassment or other more "mundane" violations of women's rights, they are also likely to shrug off violence against women during times of conflict.

5 things that the international community should do[10]

1) **Make a genuine commitment to the rights of women**. Judicial institutions are the most important mechanism to promote fundamental rights. Work to build and improve them, within

individual nations and within international organizations. Make sure the treaty establishing the International Criminal Court takes a strong stand on sexual violence, including appropriate investigations, witness protection programs, and adequate compensation for victims

2) **Make acts of violence against women during conflict an international crime.** Encourage states to enact legislation as well, to allow them to also prosecute those who commit such assaults.

3) **Change the basis for international prosecution of rape and enforced prostitution.** Now these crimes are based on violations to the honour and reputation of women. They must be seen as what they are: physical assaults and attacks on the fundamental human dignity of a person.

4) **Develop comprehensive protocols for monitoring of armed conflicts so violations of women's rights are adequately documented from the outset.** Establish guidelines and adequate funding for investigations with particular attention to issues of confidentiality, anonymity, counseling and ensuring the security of victims and witnesses

5) **Devise mechanisms which allow women to instigate civil proceedings including suits against members of the government and armed forces.** The Hague and Geneva conventions recognize this right; it would be strengthened if the international community recommended appropriate procedures for exercising it.

Notes

[1] Kofi Annan, UN Secretary-General, "By Invitation: Two concepts of sovereignty", *The Economist*, 18 September 1999.

[2] Askin, p. 277.

[3] "Rape: A Crime of War", Shelley Saywell, director, Silva Basmajian, producer, National Film Board of Canada, 1997.

[4]"The State of the World's Children Report", News Feature, p. 1, UNICEF, 1996.

[5]Askin, p. 339.

[6]Askin, p. xiii.

[7]"The World Expected Peace. It Found a New Brutality", Barbara Crossette, *The New York Times*, 1998.

[8]Neier, p. 222.

[9]Adapted from Common Ground, pp. 11–13.

These demonstrators gathered in Port-au-Prince in 1998 to protest the government's continuing failure to prosecute soldiers and police who targeted women for violence, particularly rape, during military rule in Haiti from 1991 through 1994. Failure to prosecute those who committed human rights abuses has led to anger and disillusionment for the Haitian people, and encouraged continued political instability. (Courtesy of AP/Wide World Photos)

Bosnian women from the eastern enclave of Srebrenica crying during a demonstration in Tuzla in 1996. An estimated 20,000 Bosnian women were raped, tortured and often murdered during attacks by Serbian forces in 1991–92. (Courtesy of AP/Wide World Photos)

Beatrice survived the 1994 massacre that left an estimated 1 million Rwandans dead and more than 200,000 women the victims of rape and sexual mutilation. But like many of the rape victims who survived, she is dying a slow, painful death from the AIDS virus she contracted from her attacker. (Courtesy of Magnum Photos)

An Afghan woman clad in burqa or veil begs outside a shrine in Kabul. Many widows have lost everything in the decade-long conflict in Afghanistan and have to beg if their families are to survive. Many more women, especially members of ethnic minorities, have been raped and sexually abused in order to intimidate local populations. (Courtesy of AP/Wide World Photos)

A construction crew of women and children break rocks as they widen a road outside of Rangoon, Myanmar. Forced labor is common under the current military dictatorship. Women are often sexually assaulted when they serve on work crews, as well as in their own villages after male relatives have been forced to leave and join army crews. (Courtesy of AP/Wide World Photos)

Indian troops have had a virtually free hand in subduing insurgents in the northern state of Kashmir. Rape has been a frequent weapon used by soldiers to punish women sympathetic or related to militants, and to intimidate the local population. (Courtesy of Magnum Photos)

Members of the Coalition of Indonesian Women for Justice and Democracy protest outside the headquarters of the Indonesian Defense Department in July 1998. They accused the military of not doing enough to prevent mass rapes of Indonesian Chinese women during rioting in May. More than 150 women were sexually assaulted during the riots, largely in the Chinatown section of Jakarta, Indonesia's capital city. (Courtesy of AP/Wide World Photos)

Within the first hour nearly 200 people thronged an exhibition on rape used as a weapon of terror, presented by the Singapore-based Association of Women for Action and Research (AWARE) in August 1998. The exhibition was the start of a two-systemic rapes that occurred during May 1998 anti-Chinese riots in Indonesia. (Courtesy of *The Straits Times*)

In September 1998, a delegation led by AWARE President Dr. Phyllis Chew delivered eight volumes containing 40,000 signatures to the Indonesian embassy in Singapore. This petition expressed extreme concern over attacks on Chinese Indonesian women. (Courtesy of *The Straits Times*)

An East Timorese woman cries as she remembers victims during a march in memory of women who were raped, abused or killed during militia rampages in 1999 in the capital city of Dili. There were hundreds of reports of sexual violence against women in the militia-sponsored mayhem that followed the announcement of a vote in favor of independence for the territory from Indonesia. (Courtesy of AP/Wide World Photos)

APPENDIX

Selected Bibliography

BOOKS

Allen, Beverly, *Rape Warfare: The Hidden Genocide in Bosnia-Herzegovina and Croatia,* University of Minnesota Press, Minneapolis, 1996.

Askin, Kelly Dawn, *War Crimes Against Women,* Kluwer Law International, The Hague, 1997.

Brownmiller, Susan, *Against Our Will: Men, Women and Rape,* Fawcett Columbine, Ballantine Books, 1975.

Chang, Iris, *The Rape of Nanking,* Penguin Books, London, 1997.

Hicks, George, *The Comfort Women,* Allyn & Unwin, London, 1995.

Gutman, Roy and Rieff, David, Editors, *Crimes of War: What the Public Should Know,* W.W. Norton & Company, New York and London, 1999.

Neier, Aryeh, *War Crimes,* Times Books, Random House, New York, 1998.

Sajor, Indai Lourdes, Editor, *Common Grounds: Violence Against Women in War and Armed Conflict Situations*, Asian Center for Women's Human Rights (ASCENT), Quezon City, The Philippines, 1998.

Stiglmayer, Alexandra, Editor, *Mass Rape: The War Against Women in Bosnia-Herzegovina,* University of Nebraska Press, Lincoln, 1994.

ARTICLES AND REPORTS

Amnesty International, London

"India: Summary of human rights concerns in Jammu and Kashmir", February 1995.

"Myanmar: Atrocities in the Shan State", April 1998.

"Women in Afghanistan, A Human Rights Catastrophe", March 1995.

Asian Human Rights Commission, Hong Kong

"Starvation and Militarisation in Burma", December 1999.

Human Rights Watch, New York/London/Brussels

"Human Rights Watch Global Report on Women's Human Rights", August 1995.

"India: Behind the Kashmir Conflict", July 1999.

"Indonesia: Human Rights and Pro-Independence Actions in Irian Jaya", December 1998.

"Indonesia: The Damaging Debate on Rapes of Ethnic Chinese Women", December 1998.

"Indonesia: Why Aceh is Exploding", August 1999

"Rape in Haiti: Weapon of Terror", published jointly with National Coalition for Haitian Refugees, July 1994.

"Rape in Kashmir: A Crime of War", published jointly with Physicians for Human Rights, Boston, May 1993.

"Shattered Lives: Sexual Violence during the Rwandan Genocide and Its Aftermath", September 1996.

"Thirst for Justice: A Decade of Impunity in Haiti," September 1996.

"Questions and Answers on East Timor," September 1999.

International Red Cross

Gardam, Judith, "Women, human rights and international humanitarian law", International Review of the Red Cross, pp. 421–432, September 1998.

"Women and War", August 1995.

Niarchos, Catherine, "Women, War and Rape: Challenges facing the International Tribunal for the Former Yugoslavia", pp. 674-675, *Human Rights Quarterly,* Vol. 17, No. 4, The Johns Hopkins University Press, Baltimore, November 1995.

United Nations, New York

Coomaraswamy, Radhika, "Report of the Special Rapporteur on violence against women, its causes and consequences, Addendum: Mission to Indonesia and East Timor on the issue of violence against women," January 1999.

"Report of the Special Rapporteur on violence against women, its causes and consequences, 1998.

Lallah, Rajsoomer, "Question of the violation of human rights and fundamental freedoms in any part of the world, Situation of human rights in Myanmar", January 1999.

"Women 2000, Sexual Violence and Armed Conflict: United Nations Response", Division for the Advancement of Women, April 1998.

Selected Organizations

Below is a short list of nongovernmental organizations which support the human rights of women through their research, advocacy and/or direct service. It is intended to be simply a starting point for readers ready to seek more information and opportunities to contribute.

International Groups

Amnesty International
International Secretariat
1 Easton Street
London WC1X 8 DJ
United Kingdom
www.amnesty.org
www.amnesty.se/women/beijing.htm

Human Rights Watch
350 Fifth Avenue, 34th floor
New York, NY 10118-3299
U.S.A.
www.hrw.org

International Committee of the Red Cross
19 Avenue de la Paix
CH 1202 Geneve
Switzerland
www.icrc.org

United Nations
United Nations Plaza
New York, New York 10017
U.S.A.
www.un.org
www.un.org/esa

Women for Women International
733 15th Street NW, Suite 310
Washington, DC 20005
U.S.A.
www.womenforwomen.org
info@womenforwomen.org

In Asia:

Asian Center for Women's Human Rights (ASCENT)
Suite 306 MJB Building
220 Tomas Morato Avenue
Quezon City, The Philippines
ascent@mnl.cyberspace.com.ph

Association of Women for Action and Research (AWARE)
Block 5 Dover Crescent #01-22
Singapore 130005
www.aware.org.sg

Fifteen Steps to Protect Women's Human Rights

Below is a set of important actions to protect women's human rights, recommended by Amnesty International and reprinted with its permission.

15 Point Program

Human rights for women, as for all individuals, are protected in international law. Yet women suffer the full range of human rights violations known to the modern world. Women and girl children also face human rights violations solely or primarily because of their sex.

The international community can play a decisive role in protecting human rights through vigilant and concerted action. Important steps towards protecting women's human rights worldwide include documenting human rights violations, publicizing these as widely as possible and campaigning to press government authorities for an end to the abuses. Governments which fail to protect fundamental human rights should be confronted with the full force of international condemnation.

Armed political groups should also take steps to prevent abuses of the human rights of women and girl-children.

Amnesty International's 15-point program to protect women from human rights violations contains recommendations which address both abuses primarily suffered by women, and the range of human rights violations that women have experienced along with men and children. The recommendations focus on the specific areas of Amnesty International's expertise and aim to complement and contribute to the efforts of others working on women's rights issues.

The campaign to protect women's human rights will have to be waged on the same fronts and the same issues as that to protect everyone's human rights. Some human rights violations, however, require specific action to protect women in particular. The recommendations below reflect the breadth of the campaign.

1

Governments should recognize that women's human rights are universal and indivisible

The Platform for Action to be adopted by the Fourth UN World Conference on Women must reflect the commitment made by governments in the Vienna Declaration and Programme of Action of the 1993 UN World Conference on Human Rights that "[t]he human rights of women and of the girl-child are an inalienable, integral and indivisible part of universal human rights".

2

Ratify and implement international instruments for the protection of human rights

Governments should ratify international legal instruments which provide for the protection of the human rights of women and girl-children, such as:

- the International Covenant on Civil and Political Rights (ICCPR) and its two Optional Protocols;
- the International Covenant on Economic, Social and Cultural Rights;
- the Convention against Torture and Other Cruel, Inhuman or Degrading Treatment or Punishment;
- the Convention on the Elimination of All Forms of Discrimination against Women;
- the Convention on the Rights of the Child;
- the Convention and Protocol relating to the Status of Refugees.

Governments should also ratify regional standards which protect the human rights of women and girl-children.

Governments who have already ratified these instruments should examine any limiting reservations, with a view to withdrawing them. This is particularly important in the case of the Convention on the Elimination of All Forms of Discrimination against Women, where the commitment of many governments is seriously undermined by the extent of their reservations.

Governments should take due account of non-treaty instruments such as the Vienna Declaration and Programme of Action and the Declaration on the Elimination of Violence against Women.

Governments should ensure that reports to treaty monitoring bodies include detailed information on the situation of women and girl-children.

3

Eradicate discrimination, which denies women human rights

Governments should recognize that discrimination against women, including lesbians and girl-children, is a key contributory factor to human rights abuse such as torture, including rape and other forms of custodial violence. Governments should initiate a plan of action against such discrimination.

Governments should ensure that women are treated equally in law; a woman's evidence should have the same weight as a man's in all judicial proceedings and women should not receive harsher penalties than a man would for the same offence.

Where it is alleged that discrimination in the administration of justice contributes to human rights violations against women an independent commission should be appointed to investigate and make recommendations to rectify the situation.

4

Safeguard women's human rights during armed conflict

Stop torture, including rape, "disappearances" and extrajudicial executions.

Take special steps to prevent rape during armed conflict, often the context for violent sexual abuse of women and girl-children. Bring government agents responsible for rape to justice.

The UN should ensure that personnel deployed in UN peace-keeping and other field operations observe the highest standards of humanitarian and human rights law and receive information on local cultural traditions. They should respect the rights and dignity of women at all times, both on and off duty. Human rights components of UN field operations should include experts in the area of violence against women, including rape and sexual abuse, to ensure that prisons and places of detention where women are held are clearly identified and properly investigated and that victims of rape and other custodial violence have suitable and confidential facilities to meet investigators who are specially trained and experienced in this area.

5

Stop rape, sexual abuse and other torture and ill-treatment by government agents and paramilitary auxiliaries

Take effective steps to prevent rape, sexual abuse and other torture and ill-treatment in custody.

Conduct prompt, thorough and impartial investigations into all reports of torture or ill-treatment. Any law enforcement agent responsible for such acts, or for encouraging or condoning them, should be brought to justice.

Any form of detention or imprisonment and all measures affecting the human rights of detainees or prisoners should be subject to the effective control of a judicial authority.

All detainees should have access to family members and legal counsel promptly after arrest and regularly throughout their detention and/or imprisonment.

The authorities should record the duration of any interrogation, the intervals between interrogations, and the identity of the officials conducting each interrogation and other persons present.

Female guards should be present during the interrogation of female detainees and prisoners, and should be solely responsible for carrying out any body searches of female detainees and prisoners to reduce the risk of rape and other sexual abuses. There should be no contact between male guards and female detainees and prisoners without the presence of a female guard.

Female detainees and prisoners should be held separately from male detainees and prisoners.

All detainees and prisoners should be given the opportunity to have a medical examination promptly after admission to the place of custody and regularly thereafter. They should also have the right to be examined by a doctor of their choice.

A medical examination, by a female doctor wherever possible, should be provided immediately for any woman in custody who alleges she has been raped. This is a crucial measure in obtaining evidence for legal prosecution.

Victims of rape and sexual abuse and other torture or ill-treatment in custody should be entitled to fair and adequate compensation and appropriate medical care.

6

Prevent "disappearances" and extrajudicial executions by government agents and compensate the victims

Conduct prompt, thorough and impartial investigations into all reports of "disappearances", extrajudicial executions and deaths in custody and bring to justice those responsible.

Ensure that the commission of a "disappearance" or extrajudicial execution, or causing the death of a prisoner in custody is a criminal offence, punishable by sanctions commensurate with the gravity of the practice.

Inform families immediately of any arrest and keep them informed of the whereabouts of the detainee or prisoner at all times.

Make available judicial remedies (such as habeas corpus and amparo) to enable lawyers and relatives to locate prisoners and obtain the release of anyone who has been arbitrarily detained.

Prevent detention or imprisonment other than in official, known detention centres, a list of which should be widely publicized.

Order forensic investigations into killings and deaths in custody to be carried out promptly and thoroughly by independent qualified personnel.

Provide fair and adequate redress to relatives of victims of "disappearance", extrajudicial execution and death in custody, including financial compensation.

The civil status of women whose relatives have "disappeared" should not be penalised. Identity cards, travel documents, other official papers, and state benefits should be made available to women whose relatives have "disappeared".

7

Stop persecution because of family connections

Any woman detained, imprisoned or held hostage solely because of her family connections should be immediately and unconditionally released.

The practice of killing, abducting, or torturing women in order to bring pressure on their relatives should not be tolerated. Anyone responsible for such acts should be brought to justice.

8

Safeguard the health rights of women in custody

Provide all women under any form of detention or imprisonment with adequate medical treatment, denial of which can constitute ill-treatment.

Provide all necessary pre-natal and post-natal care and treatment for women in custody and their infants.

The imprisonment of a mother and child together must never be used to inflict torture or ill-treatment on either by causing physical or mental suffering. If a child is ever separated from its mother in prison she should be immediately notified and continuously kept informed of its whereabouts and given reasonable access to the child.

Women in custody should be consulted over arrangements made for the care of their infants.

9

Release all prisoners of conscience immediately and unconditionally

Release all detainees and prisoners held because of their sex, peaceful political beliefs or activities, ethnic origin, sexual orientation, language or religion.

No woman should be detained or imprisoned for peacefully attempting to exercise basic rights and freedoms enjoyed by men.

Governments should review all legislation and practices, which result in the detention of women because of their homosexual identity or because of homosexual acts in private between consenting adults.

10

Ensure prompt and fair trials for all political prisoners

Stop unfair trials which violate the fundamental rights of political prisoners in all parts of the world.

Ensure that all political prisoners charged with a criminal offence receive a prompt and fair trial by a competent, independent and impartial tribunal.

Ensure that all political prisoners are treated in accordance with internationally recognized safeguards for fair legal proceedings.

11

Prevent human rights violations against women refugees and asylum-seekers and displaced women

No one should be forcibly returned to a country where she or he can reasonably be expected to be imprisoned as a prisoner of conscience, tortured, including by being raped, "disappeared" or executed.

Governments should remove all barriers, whether in law or administrative practice, to women seeking political asylum on the basis of persecution based on sexual identity.

Every woman refugee or asylum-seeker should be given the opportunity of an individual hearing, and should not be regarded as merely being part of her family.

Governments should take measures to protect women's physical safety and integrity by preventing torture, including rape, and illtreatment of refugee women and asylum-seekers in the country of asylum. Other forms of sexual abuse/exploitation, such as extorting sexual favours for commodities, must be prevented.

Governments should thoroughly and impartially investigate human rights violations committed against refugees and asylumseekers in the country of asylum, and bring to justice those responsible.

In procedures for the determination of refugee status governments should provide interviewers trained to be sensitive to issues of gender and culture, as well as to recognize the specific protection needs of women refugees and asylum-seekers. Those who may have suffered sexual violence should be treated with particular care, by ensuring that their cases are handled by female staff.

Women refugees and asylum-seekers should have equal access to procedures for voluntary repatriation, to ensure that those wishing to return are able do to so and to protect those who do not wish to return from refoulement.

12
Abolish the death penalty

Governments should abolish the death penalty and stop judicial executions.

All death sentences should be commuted.

Legislation which allows a woman to be put to death for an offence for which a man would receive a lesser sentence should be abolished.

In countries which retain the death penalty, the law should provide that executions will not be carried out against pregnant women and new mothers, in conformity with international standards.

13
Support the work of relevant intergovernmental and non-governmental organizations

Governments should publicly state their commitment to ensuring that the intergovernmental bodies which monitor violations of human rights suffered by women, including the UN Commission on Human Rights and its Special Rapporteur on violence against women, the UN Commission on the Status of Women and the Committee on the Elimination of Discrimination against Women (CEDAW), have adequate resources to carry out their task effectively.

The equal status and human rights of women should be integrated into the mainstream of UN system-wide activity. These issues should be regularly and systematically addressed by the relevant UN bodies and mechanisms.

Governments should guarantee that women activists and non-governmental organizations working peacefully for the promotion and protection of women's human rights enjoy all rights set out in the Universal Declaration of Human Rights and the ICCPR.

Governments participating in the Fourth UN World Conference on Women should ensure that the Platform for Action adopted at the Conference protects the fundamental civil, political, economic, social and cultural rights of women, and that the measures it contains are implemented.

14

Promote women's rights as human rights through official programs of education and training

Governments should ensure all law enforcement personnel and other government agents receive adequate training on national and international standards which protect the human rights of all women and how to enforce them properly.

Law enforcement personnel and other governments agents should be instructed that rape of women in their custody is an act of torture and will not be tolerated.

A special emphasis should be given to education designed to make women aware of their rights and to make society at large conscious of its duty to respect the human rights and fundamental freedoms of women and girl-children. Education in the human rights of women and girl-children should be integrated in all education and training policies at both national and international levels.

Special steps should be taken to uphold the UN Declaration to Eliminate Violence against Women. These steps should include a clear prohibition of gender-based violence, whether occurring in public or private life.

Governments should give high priority in development assistance projects for the implementation of human rights particularly as they affect women and girl-children. The Commission on Human Rights and its secretariat, the Centre for Human Rights, should also be encouraged to ensure that the human rights of women are always given full attention in projects carried out under the Advisory Services and Technical Assistance program. The Centre for Human Rights should be able to respond fully and promptly to requests for assistance in establishing educational programs to combat gender discrimination.

Governments and intergovernmental organizations should make available human rights education materials which promote women's rights as human rights. These materials should designed to be understood by the illiterate.

15

Armed political groups should safeguard women's human rights

Armed political groups should also take steps to prevent abuses by their members such as hostage-taking, torture, and ill-treatment, including rape, and arbitrary and deliberate killings, and to hold those responsible for such abuses to account.

~·~

EPILOGUE:
THE AWARE COMMITTEE ON RAPE

The Association of Women for Action and Research (AWARE) is a Singapore-based women's voluntary organization, founded in 1985, which offers legal advice, professional counselling and helpline assistance to women in need. AWARE's helpline service provides support for victims of rape, and sexual and physical abuse. The organization is also an advocate for change that will enable women to take their rightful places alongside men, in a just and equal society.

In 1988 the AWARE Committee on Rape was formed; members included then-president Dr. Phyllis Chew, chairwoman Doreen Liu, Janet Lyn and Ng Boon Cheng. One of the committee's first actions was to sponsor a month-long exhibition to focus public attention on the appalling use of rape as a weapon of terror throughout the world. Many of those who attended the exhibition expressed their shock, outrage and surprise at what they saw. It became clear that something more was needed to raise public awareness, in Asia and in the rest of the world. This book for the general reader seemed to be the answer.

The committee would like to thank the Lee Foundation, a long-time supporter of educational and humanitarian causes, for its financial support.